Resilient and Friendly Living Environments

The paramount goal of this series is to be a groundbreaking resource for professionals, practitioners, and enthusiasts dedicated to shaping resilient, friendly, and inclusive environments. This series aims to empower its readers with cutting-edge insights and practical tools to design spaces that seamlessly blend sustainability, adaptability, and cultural sensitivity. By delving into such topics, the series strives to provide a holistic perspective on creating environments that prioritize the well-being and interconnectedness of diverse communities. Moreover, the series also seeks to guide the development of frameworks that evolve with the ever-changing needs of societies, ensuring that the built environment remains responsive and relevant over time. Ultimately, this book series aspires to be a catalyst for positive transformation, encouraging a paradigm shift in the way we conceptualize, design, and interact with the spaces we inhabit. This series covers a broad spectrum of topics related to building environments that stand resilient and friendly, with a special emphasis on inclusivity for individuals of all ages. The series encompasses but is not limited to:

- Sustainable and Adaptive Design: Guidance on creating environments that are environmentally sustainable and capable of adapting to changing circumstances.
- Cultural Considerations in Design: Insight into designing spaces that respect and reflect the cultural diversity of communities which promote a sense of belonging among individuals.
- Supportive Infrastructure: Detailed examinations of infrastructure elements that contribute to the overall well-being, safety, and accessibility of diverse communities. It can also include wellness initiatives that address physical, mental, and emotional well-being.
- Designing for All Ages: Foster an inclusive environment that values diversity. This includes promoting equal opportunities, and ensuring that everyone feels welcome and respected. In-depth discussions on creating infrastructure and spaces that caters to the unique needs of both children and older adults is also of great importance.
- Community Engagement: Practical approaches to involving the community in the design process and ensuring that the built environment aligns with their needs and preferences.
- Safety Measures and Disaster Risk Reduction: Prioritize safety and security to ensure a comfortable and secure environment. This includes physical safety measures as well as fostering a culture of respect and mutual support.
- Technological Integration: Explorations of how technology can be seamlessly integrated to enhance connectivity, flexibility, and inclusivity within the built environment.
- Collaborative Spaces: Design work and learning environments that facilitate collaboration. This can involve open-plan workspaces, collaborative tools, and flexible seating arrangements that encourage interaction and teamwork.
- Adaptive Policies and laws: Establish policies and laws that are adaptable to changing circumstances. This flexibility allows for the accommodation of diverse needs and ensures that policies remain relevant and effective over time.

This book series is a valuable resource for architects, designers, urban planners, educators, and anyone interested in the intersection of design, community, and well-being. The series aims to inspire and guide readers in creating environments that not only withstand challenges but also flourish as vibrant and inclusive spaces for everyone.

Yibin Ao • Yan Wang • Panyu Peng
Jianjun Zhao

Building Information Modeling (BIM) Adoption

Learning, Competition, and Practice

Springer

Yibin Ao
Environment and Civil Engineering
Chengdu University of Technology
Chengdu, Sichuan, China

Panyu Peng
Chengdu University of Technology
Chengdu, Sichuan, China

Yan Wang
Engineering Management
Sichuan College of Architectural Technology
Deyang, Sichuan, China

Jianjun Zhao
Environmental and Civil Engineering
Chengdu University of Technology
Chengdu, Sichuan, China

ISSN 3005-1681 ISSN 3005-169X (electronic)
Resilient and Friendly Living Environments
ISBN 978-981-95-3992-5 ISBN 978-981-95-3993-2 (eBook)
https://doi.org/10.1007/978-981-95-3993-2

This Springer imprint is published by the registered company Springer Nature Singapore Pte Ltd.
The registered company address is: 152 Beach Road, #21-01/04 Gateway East, Singapore 189721, Singapore

Preface 1

With the rapid digital transformation of the Architecture, Engineering, and Construction (AEC) industry, Building Information Modeling (BIM) has gradually become an important tool leading industry innovation. BIM demonstrates strong advantages in project collaboration, visual modeling, and lifecycle management, and it has also driven profound changes in higher education in terms of teaching philosophy and talent training models. Universities are not only required to impart technical knowledge but also to cultivate versatile BIM professionals equipped with behavioral awareness, career cognition, and lifelong learning adaptability.

I have been teaching at Chengdu University of Technology for 15 years, giving lectures to undergraduates, master's students, and doctoral students in civil engineering, construction management, and geological engineering. Fifteen years of teaching is neither short nor long, but I can clearly feel the changes from one cohort of students to the next. Traditional lecture-based teaching methods are gradually becoming less popular, while group-based, task-based, and project-based teaching methods are more effective in stimulating students' classroom engagement. As academic competitions have become highly valued by universities, students now participate with clear motivation and strong willingness. At this point, I often reflect: does participation in academic competitions truly help students master professional knowledge and skills? How many students can persevere throughout the entire process of participating in such competitions? Can academic competitions promote college students' understanding of their major and the industry? And will these competitions help students better understand their future career directions and influence their willingness to pursue BIM-related professions? With these questions in mind, our research team has continuously tracked academic competitions and carried out a series of explorations, the results of which are presented in this book.

Therefore, *BIM Adoption: Learning, Competition, and Practice* represents the culmination of years of research achievements and practical observations by the author team in the field of BIM education. The book integrates multiple behavioral theoretical frameworks—including the Technology Acceptance Model (TAM), the Unified Theory of Acceptance and Use of Technology (UTAUT), Social Cognitive Theory (SCT), and the Theory of Planned Behavior (TPB)—to systematically examine the acceptance and practice of BIM learning among university students in both classroom and competition environments.

The book is organized into nine chapters, with a clear structure and rigorous logic. Chapter 1 reviews global trends in BIM application and related research progress, while Chap. 2 evaluates the maturity of China's BIM application policies. Chapters 3, 4, 5, 6 and 7 focus on students' learning behaviors, educational effectiveness, and competition performance, drawing on empirical data from national BIM graduation design competitions to explore key influencing factors and motivational mechanisms in BIM education. Chapter 8 further analyzes how competition experiences shape students' attention to and choices of BIM-related careers, revealing the logical pathways that connect learning, practice, and career cognition. The final chapter summarizes the findings of this book and proposes forward-looking recommendations for BIM education reform and practice.

Structurally, the book begins with a macro-level perspective, analyzing the development of China's BIM policies and the maturity of BIM technology, and then progressively moves to the micro-level, examining students' learning behaviors, motivational mechanisms, and educational outcomes. It ultimately proposes reform paths for the future, providing both theoretical support and practical guidance for building a high-quality talent system required in the digital construction era.

This book is intended for researchers, educators, and policymakers in fields such as architectural education, engineering training, education policy, and curriculum design. It also provides valuable references for industry managers and graduate students.

Finally, I would like to express my heartfelt gratitude to the students and teachers from all the universities who participated in this study, as well as to Glodon Company Limited for their contributions in data support and teaching feedback. I also extend my thanks to the Sichuan Education and Scientific Research Grant Project (Sichuan Education Letter [2023] No.478); the Sichuan Provincial Higher Education Talent Training Quality and Teaching Reform Project (2024–2026, JG2024-0600); and the Chengdu University of Technology Higher Education Talent Training Quality and Teaching Reform Project (2024–2026, JG2420006) for their funding and support of this research.

Chengdu, China
September 2025

Yibin Ao

Competing Interests The authors have no competing interests to declare that are relevant to the content of this manuscript.

Preface 2

In the sweeping tide of global digitalization, the Architecture, Engineering, and Construction (AEC) industry is undergoing a profound paradigm shift. Building Information Modeling (BIM), once merely a tool for design and visualization, has now risen to become the central engine driving efficient collaboration, smart construction, and lifecycle management across the industry. Its depth and breadth of application continue to expand, transforming BIM from an "optional" feature into both a "necessary" and "standardized" component of industry development. This fundamental transformation of technological paradigms and industry ecosystems has placed unprecedented demands on practitioners' knowledge structures, competency models, and the entire talent supply chain, while simultaneously pressing engineering education toward deep, systemic restructuring.

It is against this grand backdrop that this book has emerged. We have observed with keen awareness that the role and meaning of BIM education have fundamentally shifted. No longer confined to isolated technical training detached from professional curricula, BIM education has evolved into a critical hub that connects multiple stakeholders and empowers the future. It tightly links theoretical teaching in universities with the dynamic practices of industry, bridging the gap between traditional classroom knowledge and the complex demands of real-world projects. Increasingly, it serves as a vital bridge for students to transition smoothly from academic study to professional roles. Thus, exploring the actual effectiveness of BIM education and the driving logic behind it is not only essential for improving teaching quality but also critical for supplying the digital construction era with qualified, innovative, and competitively skilled engineering talent.

The central mission of this book is to deeply analyze the core scientific issue of *"the mechanisms influencing the effectiveness of BIM education."* We focus on the learning journeys of university students—future key actors in the industry—systematically exploring their cognitive-behavioral pathways, multidimensional learning motivations, and career transition efficacy. Of particular significance is the book's emphasis on competition-based learning activities as a major entry point of research. We investigate how such highly practical, challenging, and socially interactive activities can be effectively embedded within the BIM education system, and how their unique organizational forms, incentive mechanisms, and evaluative feedback processes generate "catalytic effects" in enhancing students' technical mastery, innovative thinking, teamwork, and capacity to address complex problems.

To build a solid and reliable analytical framework, we integrate multiple classical theoretical models known for their strong explanatory power. These include the Technology Acceptance Model (TAM), the Unified Theory of Acceptance and Use of Technology (UTAUT), Social Cognitive Theory (SCT), and the Theory of Planned Behavior (TPB). Through cross-theoretical integration and empirical validation, this book aims to construct a multidimensional evaluation framework centered on the student experience, incorporating cognition, motivation, behavior, and social context. This framework is designed not only to identify the key factors influencing BIM learning outcomes—such as teacher competency, curriculum design, practical resources, technical support, industry linkages, and individual traits—but also to illuminate the complex interactions among these factors and the deeper mechanisms driving student learning behaviors.

The purpose of this book extends beyond theoretical exploration. It aspires to provide empirical support and actionable recommendations for the advancement of BIM education. We hope this research will serve as a reference for instructional designers, education policymakers, and industry training institutions, while also inspiring broader discussions on *how higher education talent training systems must transform in the era of digital construction.*

Chengdu, China
September 2025

Yan Wang

Preface 3

"Digital construction" is not merely a technological revolution; it represents a systemic reshaping of educational philosophy, talent structures, and industrial ecosystems.

My encounter with Building Information Modeling (BIM) dates back to my undergraduate years, when I first stepped into a computer lab and learned how to create three-dimensional models with software and generate drawings in a virtual environment. At first, it felt novel, but I soon realized that the significance of BIM extends far beyond operational skills. It is not simply a new way of drafting; it embodies a data-driven mindset, a logic of cross-disciplinary collaboration, and an inevitable trend driving the construction industry toward digitalization and intelligent development.

This realization led me to ask deeper questions: What kinds of psychological and behavioral changes do students undergo in the process of learning BIM? Do they merely "learn the software" to complete classroom tasks, or do they gradually form new perceptions and attitudes toward the future of the construction industry? When they participate in intensive, task-driven BIM competitions, can such informal learning contexts further stimulate their motivation—or even shape their future career choices?

With these questions in mind, I began to conduct systematic research. My focus gradually shifted from the classroom to the competition arena, from learning intentions to behavioral modeling, and from mere tool operation to the complex mechanisms of motivation and capability growth. I sought to interpret students' learning trajectories through frameworks such as TAM, UTAUT, and TPB. I visited universities, engaged in face-to-face conversations with teachers and students, and listened to their learning experiences and challenges. I also observed competition settings, noting how teams divided roles, advanced tasks, and presented outcomes, striving to capture the educational significance behind the data. Over time, I came to realize that BIM education is not only about teaching a course; it is about how students build confidence, develop professional identity, and transition into their careers.

This book is therefore not simply a compilation of research results, but the crystallization of years of questioning and exploration. It focuses on three essential "connections":

1. The connection between the dissemination of BIM technology and student cognition: how technology moves from passive acceptance to active understanding.
2. The connection between teaching processes and learning motivation: how classroom instruction, practice tasks, and inner drives interact.
3. The connection between learning experiences and career choices: how classroom and competition experiences shape students' confidence and interest in future careers.

Grounded in the dual pathways of "classroom + competition," and drawing on large-scale data from the National BIM Graduation Design Competition, this book develops explanatory models that progress from learning behaviors and motivational mechanisms to educational outcomes and career orientations, and empirically validates their effectiveness. It seeks to answer not only *what* students learn, but also *why* they learn, *how* they learn, and *what* they do afterward. This shift from "tool use" to "behavioral understanding" is precisely the perspective that BIM education urgently needs, and a necessary condition for cultivating versatile talent in the era of digital construction.

I am well aware that this book cannot cover every issue in BIM education, nor can it resolve all the challenges of practice. But I hope it offers researchers, educators, and industry practitioners a reference framework to reflect on how to better teach BIM, learn BIM, and cultivate BIM professionals who truly meet the demands of the future. If it sparks interdisciplinary discussion and fosters closer alignment between education and industry, then its purpose will have been fulfilled.

Finally, I would like to thank all the students and teachers who participated in the surveys—their candid sharing and practical experiences provided the most solid foundation for this research. I am also deeply grateful to my project team for their sustained collaboration over the years, and to the National Natural Science Foundation of China for its long-term support. Without these efforts, this book would not have been possible.

In the great tide of digital construction, I will continue to pursue the power of education, explore pathways for talent growth, and join hands with colleagues to envision the future of BIM education and industry development.

Chengdu, China
September 2025

Panyu Peng

Preface 4

When we talk about **Building Information Modeling (BIM)**, it is often framed as a technological innovation. Yet in reality, BIM has long surpassed the boundaries of technology, becoming a critical lever for driving the digital transformation of the construction industry.

Across the globe, more and more countries have integrated BIM into engineering standards and public project workflows. The collaborative models, information precision, and construction efficiency it enables are profoundly reshaping the entire AEC (Architecture, Engineering, and Construction) industry. Behind this sweeping transformation, however, one fundamental question continues to resurface, with few systematic answers: Are we truly prepared to cultivate the next generation of BIM professionals?

Over the past decade, countless universities have established BIM courses, added training modules, and introduced certification systems. But what have these efforts achieved in practice? Have students genuinely mastered the core logic of BIM? Are current educational formats aligned with the needs of industry? Do factors such as teachers' understanding, students' motivation, competition participation, and curriculum design form identifiable and explainable relationships?

This book was written in response to these pressing questions. Taking the chain of "learning—competition—practice" as its entry point, it combines theoretical modeling with empirical research to construct a comprehensive framework for understanding the behavioral mechanisms of BIM education. It systematically addresses three core issues:

1. The behavioral characteristics, motivational patterns, and levels of acceptance among university students in BIM learning;
2. The mechanisms through which BIM competitions enhance learning performance, teamwork, and skills development;
3. How educational experiences shape students' attention to and choices of BIM-related career pathways.

Drawing on perspectives from education, construction management, and behavioral psychology, and using data from multiple Chinese universities as well as national-level competition platforms, this book blends policy context analysis with behavioral path modeling. Its aim is to provide both an insightful analytical framework and practical recommendations for advancing BIM education reform.

We believe that the essence of BIM education is not merely to "teach tools," but to "develop capabilities" and, more importantly, to "inspire futures." If BIM is the key to intelligent construction, then education is the hand that determines whether that key can truly unlock the door.

Chengdu, China
September 2025

Jianjun Zhao

Contents

Chapter 1
BIM Adoption: A Review of Global Trends, Research Progress, and Future Directions

Abstract Building Information Modeling (BIM), as a core technology in the digital transformation of the construction industry, has been continuously expanding and deepening worldwide. This paper systematically reviews the evolution of BIM-related policies, standardization efforts, and industrial integration pathways in different countries, highlighting that its promotion generally exhibits a composite pattern of "government guidance—standard formulation—market adaptation." At the academic level, BIM research themes have evolved from an early focus on technical functionalities to encompass adoption behaviors, educational models, and project performance, with theoretical foundations primarily drawn from behavioral models such as TAM, UTAUT, and TPB. This evolution reflects a shift from qualitative to quantitative approaches, from organizational to individual perspectives, and from macro- to micro-level analyses. However, significant gaps remain in the field of BIM education, such as modeling student learning pathways, evaluating competition-based learning mechanisms, and developing interdisciplinary curriculum systems. Based on a review of global development trends and research progress, this paper proposes that future work should strengthen educational behavior modeling, develop robust assessment tools, and investigate region-specific adaptation mechanisms, in order to further advance BIM education theory and support its effective implementation in practice.

Keywords Building Information Modeling (BIM) · TAM · UTAUT · Digital construction · BIM education

1.1 Introduction

As the construction industry enters the deep waters of digital transformation, Building Information Modeling (BIM), as an integrated system for design, collaborative construction, and full life-cycle management, is profoundly reshaping the production and organizational logic of the global Architecture, Engineering, and

Y. Ao et al., *Building Information Modeling (BIM) Adoption*, Resilient and Friendly Living Environments, https://doi.org/10.1007/978-981-95-3993-2_1

Construction (AEC) industry. BIM not only enhances information integration and engineering efficiency but also plays a critical role in sustainable construction, intelligent operations and maintenance, and full-process project control (Xue et al., 2021; Zhang et al., 2024). Its core value has been widely recognized by governments and industry organizations worldwide and has been gradually incorporated into policy frameworks, industry standards, and educational curricula.

In terms of international adoption pathways, the UK government proposed the Government Construction Strategy in 2011, which required the mandatory implementation of BIM Level 2 for all government-funded projects from 2016 onward, triggering a wave of BIM standardization across Europe (Infrastructure & Authority, 2016). Singapore, through its Building and Construction Authority (BCA), introduced a BIM Roadmap in 2010, along with targeted training programs and an approval system to support its implementation (BCA, 2013). In China, during the 14th Five-Year Plan period, BIM was designated as a core technology for digital construction, with pilot projects established in government-invested engineering works across multiple provinces and municipalities (MOHURD, 2022).

At the same time, global research interest in BIM has grown rapidly. Initial studies focusing on technical functionalities have gradually expanded to encompass organizational adoption behaviors, impacts on project performance, and innovations in teaching approaches. The Technology Acceptance Model (TAM) and the Unified Theory of Acceptance and Use of Technology (UTAUT) have been widely applied to analyze individual and organizational behavioral characteristics in BIM adoption (Venkatesh et al., 2003). In the field of education, Adamu and Thorpe (2016) reviewed BIM teaching practices and highlighted systemic shortcomings in curriculum design, tool integration, and faculty capability. Succar (2010) proposed the BIM Maturity Matrix, emphasizing that BIM education should align with industry standards and competency frameworks (Succar et al., 2013).

However, current research on BIM education remains relatively fragmented, with insufficient systematic investigation into student behavioral mechanisms, competition-based learning models, and the career-oriented impacts of education. Problems such as unclear curriculum integration pathways, underdeveloped educational evaluation systems, and a disconnect between teaching methods and industry practices still persist.

Therefore, to further clarify the policy logic, industrial mechanisms, and educational transformation pathways of BIM promotion, this chapter presents a systematic review of global BIM application trends, research progress, and future challenges. Through literature assessment and problem synthesis, it identifies the theoretical foundations and evolutionary trajectories of BIM education research, laying a conceptual and methodological framework for the empirical studies in subsequent chapters.

1.2 Global Promotion Trends

1.2.1 Policy-Driven Mechanisms

The widespread adoption of BIM technology depends not only on industry-led innovation but is also profoundly shaped by government policy initiatives and institutional arrangements. Globally, many countries have employed top-down strategic planning to integrate BIM into their construction modernization agendas, making its use mandatory or prioritized in government-funded projects to create a "point-to-surface" spillover effect.

The United Kingdom is among the earliest countries to promote BIM adoption through policy intervention. In 2011, the UK government released the Government Construction Strategy, positioning BIM as a key tool to reduce costs in public construction projects. By 2016, it had formally implemented the BIM Level 2 requirement, mandating that all central government-funded projects submit BIM Level 2-compliant deliverables at the tender stage (Infrastructure & Authority, 2016). To support this mandate, the UK introduced the PAS 1192 series of standards, which were later consolidated into the ISO 19650 international standard series, setting a global benchmark for BIM standardization.

Singapore, through its Building and Construction Authority (BCA), developed a comprehensive BIM Roadmap and, since 2010, has promoted the full-process application of BIM across building approval, collaborative design, and construction supervision. From 2015 onward, the government required that projects with a gross floor area exceeding 5000 m^2 submit BIM models at the approval stage. This mandate was coupled with financial subsidies, certification systems, and professional training programs (BCA, 2013). Such a "mandatory enforcement + incentive support" approach has effectively accelerated BIM industrialization and encouraged regional enterprises and universities to increase their BIM-related investments.

Nordic countries such as Norway (Statsbygg), Finland (Senate Properties), and Denmark (Digital Construction) have also leveraged national public construction agencies to establish BIM application frameworks and mandatory technical requirements. For example, since 2007, Finland's national property client Senate Properties has required that all design projects submit BIM data files in the IFC standard format, emphasizing openness and reusability of model data (Hussein, 2021; Wong et al., 2010).

In the United States, although there is no unified federal mandate, multiple state governments (e.g., Texas Department of Transportation) and federal agencies (e.g., U.S. General Services Administration, GSA) have advanced BIM pilot applications in public works. Additionally, the National Institute of Building Sciences (NIBS) and buildingSMART USA have issued a range of technical guidelines and maturity assessment standards, forming a "standards-led application" model for BIM promotion (Suermann & Issa, 2010; Zhu, 2024).

In recent years, the Chinese government has accelerated its BIM policy rollout. The 14th Five-Year Plan for the Development of the Construction Industry explicitly proposed "accelerating the application of Building Information Modeling and promoting the digital transformation of the construction industry," setting the goal of achieving widespread BIM adoption in government-invested projects by 2025 (MHURD, 2022). Various provinces and cities, including Shanghai, Shenzhen, Sichuan, and Hubei, have subsequently issued local technical guidelines and implementation measures to apply BIM in key areas such as rail transit, affordable housing, and water conservancy projects. Meanwhile, the Ministry of Housing and Urban-Rural Development (MOHURD) has collaborated with universities and research institutions to promote the compilation of BIM teaching materials, curriculum accreditation, and industry skill standards, aiming to bridge the gap between policy implementation, industry uptake, and talent cultivation.

In summary, BIM adoption has generally followed a "policy-guided diffusion path": governments first pilot its use in public projects, then gradually expand industry-wide adoption through supporting measures such as standard formulation, financial subsidies, regulatory reform, and talent certification systems. Throughout this process, policy continuity, enforcement strength, and coordination mechanisms are key determinants of promotion effectiveness.

1.2.2 Standard Systems and Industry Integration

The establishment of standard systems and technological integration capabilities forms the foundation for the industrialization and large-scale application of BIM technology. To improve data consistency and traceability in cross-phase and cross-disciplinary collaboration, the International Organization for Standardization (ISO) released the ISO 19650 series—core standards for BIM information management—in 2018. These standards cover key aspects such as information exchange, role responsibilities, and delivery stage definitions, providing a unified framework for global BIM life-cycle collaboration (Almeida et al., 2023; Standardization, 2022).

The ISO 19650 series, derived from the UK's PAS 1192 standards, is currently the most widely adopted information management standard globally, embraced in regions across Europe, Asia, and the Middle East. Its main components include structured specifications for Exchange Information Requirements (EIR), Master Information Delivery Plans (MIDP), and Asset Information Models (AIM); emphasis on standardized naming conventions, file transfer mechanisms, and clearly defined responsibilities; and encouragement of open data formats (e.g., IFC, COBie) to enable interoperability and mutual recognition between platforms.

In China, the Ministry of Housing and Urban-Rural Development (MOHURD) issued the Unified Standard for Building Information Modeling Application (GB/T 51212-2016) and has progressively aligned it with ISO 19650. However, as BIM adoption is still in a developmental stage, the standard system has not yet fully integrated design, construction, and operation phases, resulting in fragmentation and inconsistencies in technical specifications.

From a technological development perspective, BIM has evolved from being an isolated modeling tool to deeply integrating with other digital technologies. Typical integration pathways include:

BIM + GIS: Linking above- and below-ground, as well as building- and district-scale data, to support city-level spatial decision-making (Wang et al., 2019).
BIM + IoT: Using sensors and real-time data collection to enhance model dynamism, enabling intelligent monitoring of construction sites and optimized operations and maintenance (Desogus et al., 2021).
BIM + Digital Twin: Creating a mirrored mapping between physical and virtual spaces to improve facility performance throughout its life cycle, with applications in smart campuses, intelligent hospitals, and rail transit systems (Pavón et al., 2025).

In addition, emerging technologies such as Artificial Intelligence (AI), Blockchain, and Augmented/Virtual Reality (AR/VR) are being integrated into BIM platforms, accelerating its transformation from a "visual modeling tool" to an "intelligent decision-making platform" (Pan & Zhang, 2023).

Despite this clear trend toward integration, fragmented standards and the complexity of coordination remain major obstacles to widespread BIM adoption. On one hand, the closed ecosystems of software vendors lead to inconsistent data format standards and poor model interoperability, restricting information exchange among project stakeholders. Although the Open BIM movement promotes the adoption of open standards such as IFC and gbXML, real-world projects still face compatibility, accuracy, and data loss issues (Jiang et al., 2019).

On the other hand, the pace of BIM standard development varies significantly across countries and regions. In developing countries and among small- and medium-sized enterprises (SMEs), the high cost of standard implementation, steep technical learning curves, and long return-on-investment periods often slow progress (Olugboyega, 2024). This exacerbates the "digital divide" in BIM adoption and hinders its sustainable global expansion.

Therefore, driving the deep integration of BIM with industry requires the creation of open, interoperable, and practical standard systems, along with extending these standards into educational content, professional certification, and performance evaluation frameworks. At the same time, governments and industry bodies should introduce more operational measures—such as translating standards into actionable guidelines, providing cost subsidies, and establishing SME support mechanisms—to ensure the transition "from standards to practice" becomes a reality.

1.3 Academic Research Progress

The continuous evolution of BIM technology has not only driven transformative changes in construction industry practices but has also inspired a diverse range of academic research topics at the theoretical level. These studies cover multiple domains, including technology adoption, education and training, collaboration mechanisms, and performance evaluation, and exhibit a trend of ongoing expansion and deepening.

1.3.1 Research Themes and Theoretical Foundations

Based on existing literature, BIM-related research can be broadly classified into the following four thematic areas, each of which has developed relatively mature theoretical underpinnings:

1. Technology Acceptance and Organizational Adoption

   This is one of the most fundamental and widely studied directions in BIM research, focusing on the adoption behavior of BIM technology at the levels of project teams, organizations, and even entire industries. Many scholars employ the Technology Acceptance Model (TAM) and the Unified Theory of Acceptance and Use of Technology (UTAUT) as analytical frameworks to examine how individuals or organizations accept BIM tools, processes, and collaborative workflows, as well as the factors influencing such acceptance (Venkatesh et al., 2003). For example, Park et al. (2019), through a questionnaire survey of Korean construction firms, found that performance expectancy, social influence, and organizational support play significant roles in the BIM adoption process.
2. BIM Education and Skills Development

   As BIM has been increasingly incorporated into architecture- and engineering-related curricula in higher education institutions, the educational dimension has become a prominent research focus. The emergence of BIM competitions, cross-disciplinary practical training, and immersive teaching methods has fostered new educational models that continue to evolve.
3. Project Performance and Collaborative Innovation

   BIM has demonstrated substantial benefits in improving project quality, optimizing resource allocation, and reducing construction conflicts. Accordingly, relevant research often concentrates on the impact of BIM on collaborative efficiency during the construction phase, schedule control, and project cost management. For example, Eldeep et al. (2022), through case analysis, argued that BIM facilitates early conflict detection, optimizes material planning, and enhances on-site management collaboration, making it a key technology for advancing lean construction practices.
4. Educational Effectiveness and Behavioral Mechanism Analysis

   With the introduction of behavioral science and educational evaluation methods, an increasing number of studies have begun to focus on the mechanisms underlying BIM education outcomes. Ahankoob et al. (2025), building on the BIM Maturity Matrix, proposed a multi-level learning pathway framework, emphasizing that BIM education should go beyond software skills to include understanding collaborative logic, information management, and data-oriented thinking. Furthermore, factors such as learning motivation, self-efficacy, and collaborative abilities in BIM education have gradually become focal points, creating a new interdisciplinary area at the intersection of educational psychology and engineering education.

The sustained development of these research themes has laid a solid foundation for the construction of the BIM theoretical system and for educational reform, while also driving the continuous evolution of research methodologies.

1.3.2 Regional Distribution and the Evolution of Research Methods

In terms of regional distribution, BIM-related research was initially dominated by North American and Western European countries, mainly concentrated in regions with highly modernized construction industries, such as the United States, Canada, the United Kingdom, Finland, and Germany. These countries not only possess a mature construction industry foundation but also established BIM promotion policies and technical standards at an early stage, fostering an innovative environment driven by government–industry–academia–research collaboration. For instance, the National Institute of Building Sciences (NIBS) in the United States issued the National BIM Standard–United States (NBIMS-US), providing a general framework for BIM application. The United Kingdom, through the PAS 1192 standards and the subsequent ISO 19650 series, has led global BIM standardization efforts (Ajayi et al., 2023; Dadmehr & Coates, 2019).

In recent years, however, emerging economies have shown strong momentum in BIM research. China, for example, has seen its government frequently release BIM promotion policies and advance the localization of technical standards, spurring a large volume of applied research and higher education reform initiatives. Under the leadership of the Ministry of Housing and Urban-Rural Development, China issued the Unified Standard for Application of Building Information Modeling along with various industry guidelines, while designating national BIM education model universities. Similarly, India and Brazil have promoted the integration of BIM with local construction industry practices in fields such as software development and civil engineering education, forming regionally adapted educational models and technical pathways. Compared with developed countries, these nations' research tends to emphasize practicality, localization, and policy effects, highlighting BIM's strategic role and educational guidance function in the evolution of their construction markets.

From a methodological perspective, BIM research has undergone a clear shift from a predominance of qualitative approaches to a stronger emphasis on quantitative analysis. Early studies often relied on qualitative methods such as case studies, literature reviews, and expert interviews, mainly for exploratory problem definition and best practice identification. For example, Chahrour et al. (2021) summarized the economic and technical value of BIM through representative project case studies, emphasizing its capabilities in collaboration and clash detection.

With the standardization of research paradigms, more scholars have begun to employ Structural Equation Modeling (SEM) to reveal the underlying path mechanisms of BIM adoption behavior (Ma et al., 2020; Wu et al., 2021). In addition, methods such as Fuzzy Analytic Hierarchy Process (FAHP), Analytic Hierarchy Process (AHP), System Dynamics (SD), and Multi-Objective Optimization (MOO) have been applied to topics such as BIM education course evaluation, standard adaptability assessment, and implementation pathway selection.

It is worth noting that BIM education research methods are evolving toward a dual emphasis on data-driven approaches and behavioral modeling. On one hand, researchers are integrating multi-source datasets—such as course grades, online learning logs, and competition records—and applying clustering analysis, path regression, and other techniques to map students' learning behavior trajectories. On the other hand, scenario-based experimental design and gamified simulation are increasingly being introduced into BIM teaching contexts to evaluate learning motivation, self-directed learning capacity, and collaborative performance. For instance, Zaker and Coloma (2018) designed a BIM collaboration task based on a virtual reality platform to explore the interaction between team cognitive differences and technology acceptance.

Moreover, driven by trends in digital construction and multi-technology integration, BIM research is expanding toward interdisciplinary and system integration directions. In the future, integrating advanced methods such as machine learning, agent-based modeling (ABM), and complex network analysis into educational research will be critical to deepening BIM education studies.

1.4 Research Gaps and Future Directions

Although research on BIM technology has made remarkable progress over the past decade, establishing an initial theoretical foundation and application framework in areas such as construction project management, integrated design–construction workflows, collaborative mechanisms, and educational transformation, several critical gaps remain in terms of systematization, depth, and interdisciplinary integration. These gaps continue to limit the potential for further theoretical development and practical translation.

1.4.1 Key Research Gaps

Despite the rapid growth of BIM education research in recent years, studies targeting student-oriented applications still face four urgent challenges—also the focal points of this book's empirical investigations:

1. Lack of systematic modeling of student learning motivation and behavior. Most existing studies apply classical technology adoption models such as TAM and UTAUT to analyze students' acceptance of BIM technology, focusing primarily on the intention dimension. However, they lack in-depth characterization of behavioral trajectories, participation drivers, and strategy selection throughout the learning process. In particular, research on university students' learning pathways seldom integrates cognitive factors with actual behavioral performance to form a comprehensive mechanism analysis.
2. Insufficient research on the educational value of competition-based learning contexts. With the rise of BIM graduation design competitions and nationwide student modeling contests, competitions are increasingly becoming a crucial bridge between classroom learning and professional development. Nevertheless, current studies provide insufficient exploration of role allocation, team dynamics, and skill development pathways within competitions, and they lack systematic analysis of the linkages between competition participation and career focus.
3. Incomplete educational evaluation indicator systems. Most BIM education effectiveness studies still rely on self-perception surveys and subjective interviews, lacking objective measurement tools for dimensions such as modeling skills, collaboration awareness, and information communication. Moreover, there is no integrated evaluation framework that applies across different teaching contexts—courses, competitions, and project-based practice.
4. Lack of localized educational mechanism research for the Chinese higher education context. Current BIM education research often draws on European and American models, focusing on generic framework construction, but failing to adequately reflect the structural challenges in Chinese universities, such as "limited curriculum resources—uneven teacher competency—weak school–enterprise collaboration." Cultural differences in learning motivation, competition attitudes, and career expectations have also not been incorporated into existing analytical models.

1.4.2 Suggested Future Research Directions

To address these gaps and promote the evolution of BIM education from "tool introduction" toward "competence building" and "career orientation," this book focuses on Chinese undergraduate students in architecture, engineering, and construction disciplines, conducting systematic empirical exploration across multiple dimensions and proposing the following research pathways:

1. Detailed micro-level modeling of student behavioral mechanisms and competition motivation. Through surveys and comparative analyses of students' BIM learning and competition experiences, future research should investigate the interactive mechanisms among motivation level, self-efficacy, and professional identity, as well as the role of competition-based learning in fostering intrinsic motivation and building professional confidence.

2. Constructing a competition-context-based learning effectiveness model. Drawing on national BIM graduation design competition data, empirical modeling can be used to analyze the pathways linking team structure, role allocation, competition performance, and students' BIM competencies, thereby exploring the "competition–competence–career orientation" mechanism and providing a theoretical basis for evaluating informal learning pathways.
3. Exploring the impact mechanism of educational pathways on career focus. Future work should establish explanatory models that include learning experiences, competition participation, course evaluations, and industry awareness to reveal how BIM education shapes students' willingness to enter the industry through cognitive construction.
4. Promoting localization of teaching models and regional adaptation research. Based on the teaching environment and faculty structure of Chinese universities, research should analyze the heterogeneity of BIM learning outcomes under different teaching resource conditions and propose structural improvement strategies for BIM teaching in higher education, offering references for regional education policy-making.

In conclusion, through eight empirical chapters, this book follows the main thread of "learning motivation—competition process—competence development—career orientation," systematically addressing the current weaknesses in educational behavior modeling, the lack of competition research, and the absence of evaluation standards. It aims to shift BIM education research from fragmented observation toward structured analysis, providing empirical support and strategic guidance for cultivating digital construction talent in China.

References

Adamu, Z., & Thorpe, T. (2016). How universities are teaching BIM: A review and case study from the UK. *Journal of Information Technology in Construction, 21*, 119–139.

Ahankoob, A., Abbasnejad, B., & Aranda-Mena, G. (2025). Building information modelling (BIM) acceptance and learning experiences in undergraduate construction education: A technology acceptance model (TAM) perspective – An Australian case study. *Buildings, 15*(11), 1804.

Ajayi, S. O., Oyebiyi, F., & Alaka, H. A. (2023). Facilitating compliance with BIM ISO 19650 naming convention through automation. *Journal of Engineering, Design and Technology, 21*(1), 108–129.

Almeida, R., Chaves, L., Silva, M., Carvalho, M., & Caldas, L. (2023). Integration between BIM and EPDs: Evaluation of the main difficulties and proposal of a framework based on ISO 19650: 2018. *Journal of Building Engineering, 68*, 106091.

BCA. (2013). *BIM essential guide for BIM adoption in an organization*. MND Complex Singapore 069110.

Chahrour, R., Hafeez, M. A., Ahmad, A. M., Sulieman, H. I., Dawood, H., Rodriguez-Trejo, S., … Dawood, N. (2021). Cost-benefit analysis of BIM-enabled design clash detection and resolution. Construction Management and Economics, 39(1), 55–72.

Dadmehr, N., & Coates, S. (2019). An approach to "National Annex to ISO 19650-2." In *14th international postgraduate research conference*.

Desogus, G., Quaquero, E., Rubiu, G., Gatto, G., & Perra, C. (2021). Bim and iot sensors integration: A framework for consumption and indoor conditions data monitoring of existing buildings. *Sustainability, 13*(8), 4496.
Eldeep, A. M., Farag, M. A., & Abd El-hafez, L. (2022). Using BIM as a lean management tool in construction processes – A case study. *Ain Shams Engineering Journal, 13*(2), 101556.
Hussein, A. (2021). Utilizing life cycle assessment (LCA) during building design using building information modelling (BIM): Case studies from Finland.
Infrastructure, & Authority, P. (2016). Government construction strategy 2016–2020. In: Infrastructure and projects authority. London.
Jiang, S., Jiang, L., Han, Y., Wu, Z., & Wang, N. (2019). OpenBIM: An enabling solution for information interoperability. *Applied Sciences, 9*(24), 5358.
Ma, P., Zhang, S., Hua, Y., & Zhang, J. (2020). Behavioral perspective on BIM postadoption in construction organizations. *Journal of Management in Engineering, 36*(1), 04019036.
MHURD. (2022). Notice of the ministry of housing and urban-rural development on printing and distributing the development plan of the construction industry during the 14th five-year plan.
MOHURD. (2022). Notice of the ministry of housing and urban-rural development on printing and distributing the development plan of the construction industry during the 14th five-year plan.
Olugboyega, O. (2024). Differential relationships in the BIM implementation process in a developing country: the role of essential BIM implementation strategies. *Engineering Construction and Architectural Management, 31*(8), 3283–3307.
Pan, Y., & Zhang, L. (2023). Integrating BIM and AI for smart construction management: Current status and future directions. *Archives of Computational Methods in Engineering, 30*(2), 1081–1110.
Park, E., Kwon, S. J., & Han, J. (2019). Antecedents of the adoption of building information modeling technology in Korea. *Engineering Construction and Architectural Management, 26*(8), 1735–1749.
Pavón, R. M., Alberti, M. G., Álvarez, A. A. A., & Cepa, J. J. (2025). Bim-based digital twin development for university campus management. Case study ETSICCP. *Expert Systems with Applications, 262*, 125696.
Standardization, I. O. f. (2022). *Organization and digitization of information about buildings and civil engineering works, including building information modelling (BIM)-information management using building information modelling: Organisation et Numérisation des Informations relatives aux Bâtiments et Ouvrages de Génie civil, Y Compris Modélisation des Informations de la construction (BIM)-Gestion de L'information par la Modélisation des Informations de la construction.* Information Exchange. Échange D'informations. ISO.
Succar, B. (2010). Building information modelling maturity matrix. In *Handbook of research on building information modeling and construction informatics: Concepts and technologies* (pp. 65–103). IGI Global Scientific Publishing.
Succar, B., Sher, W., & Williams, A. (2013). An integrated approach to BIM competency assessment, acquisition and application. *Automation in Construction, 35*, 174–189.
Suermann, P. C., & Issa, R. R. (2010). The US national building information modeling standard. In *Handbook of research on building information modeling and construction informatics: Concepts and technologies* (pp. 138–154). IGI Global Scientific Publishing.
Venkatesh, V., Morris, M. G., Davis, G. B., & Davis, F. D. (2003). User acceptance of information technology: Toward a unified view. *MIS Quarterly, 27*(3), 425–478. https://doi.org/10.2307/30036540
Wang, H., Pan, Y., & Luo, X. (2019). Integration of BIM and GIS in sustainable built environment: A review and bibliometric analysis. *Automation in Construction, 103*, 41–52.
Wong, A. K., Wong, F. K., & Nadeem, A. (2010). Attributes of building information modelling implementations in various countries. *Architectural Engineering and Design Management, 6*(4), 288–302.
Wu, Z., Jiang, M., Li, H., Luo, X., & Li, X. (2021). Investigating the critical factors of professionals' BIM adoption behavior based on the theory of planned behavior. *International Journal of Environmental Research and Public Health, 18*(6), 3022.

Xue, K., Hossain, M. U., Liu, M., Ma, M., Zhang, Y., Hu, M., ... Cao, G. (2021). BIM integrated LCA for promoting circular economy towards sustainable construction: an analytical review. *Sustainability, 13*(3), 1310.

Zaker, R., & Coloma, E. (2018). Virtual reality-integrated workflow in BIM-enabled projects collaboration and design review: a case study. *Visualization in Engineering, 6*(1), 4.

Zhang, S., Zhang, S., Wang, C., Zhu, G., Liu, H., & Wang, X. (2024). Extended IFC-based information exchange for construction management of roller-compacted concrete dam. *Automation in Construction, 163*, 105427.

Zhu, X. (2024). *Smart compliance checking frameworks for BIM standards*. Cardiff University.

Chapter 2
Building Information Modeling Learning Behavior of AEC Undergraduate Students in China

Abstract With the popularization and application of Building Information Modeling (BIM), the demand for BIM technical talents in the construction industry is increasing. Exploring college students' BIM technical learning behavior is of great practical significance to improve education activities. Based on the Unified Theory of Acceptance and Use of Technology (UTAUT), this research adds learning attitude variables to construct a theoretical model of influencing factors of college students' BIM technology learning behavior. Chinese undergraduate students were asked to complete online questionnaires through peer-to-peer contact with sample colleges and universities. Finally, 1090 valid questionnaires were obtained. The students were sampled from research-oriented, applied research-oriented, application-oriented, and private research-oriented universities in seven regions of China: northeast, north, east, south, central, northwest, and southwest. The structural equation model was used to analyze the sampling data. The results indicate that college students' BIM learning attitude, performance expectations, and social influence positively and directly impact their learning intention, which indirectly impacts their learning behavior. At the same time, promoting factors and learning intention demonstrate a significant positive and direct impact on learning behavior. Therefore, the following suggestions have been put forward to enhance college students' learning motivation for BIM technology: increase the popularization of BIM technology in colleges and universities and improve the operation level of full-time BIM teachers. The latter enables colleges and universities to continuously and stably export qualified BIM technical talents to society and the market, resulting in a continuous industry development cycle.

Keywords Learning behavior · BIM · Influencing factors · UTAUT theoretical model · Structural equation modeling

Y. Ao et al., *Building Information Modeling (BIM) Adoption*, Resilient and Friendly Living Environments, https://doi.org/10.1007/978-981-95-3993-2_2

2.1 Introduction

The construction industry has recently adopted Building Information Modeling (BIM). Moreover, research and education on BIM technology have become a prevalent topic for builders and educators in the construction industry (Ao et al., 2021; Yan et al., 2021). Relevant research shows that the popularization and application of BIM technology have undergone a significant transformation in the architecture, engineering, and construction (AEC) industry, which has increased the employment competition of fresh graduates of those majors (Wu & Issa, 2014). Moreover, college students' mastery of BIM technology can promote the application of BIM technology in the whole construction industry and make the industry more efficient (Ao et al., 2021). Such a reinforcing circle should be strengthened to accelerate the industry transformation. Therefore, it is necessary to understand the current situation of BIM technology learning behavior among college students and analyze the significant positive influence path of such behavior so that better education can be provided to stimulate the reinforcing circle.

In the BIM application and education domain, scholars mainly focus on correlating industry, project, and education aspects. For example, Ebrahimi et al. (2019) seek the interactions of sustainability and BIM in support of existing buildings. Eadie et al. (2013) conducted a survey on the use of BIM in the project life cycle, showing that the BIM industry lacks professional knowledge and training, and relevant BIM educators get more opportunities. Moreover, Tsai et al. (2019) developed a new online BIM learning course, focusing on improving the BIM education mode. Jia (2021) compared the BIM courses in Chinese and American universities and suggested that the BIM courses in China should be integrated with traditional courses. Meanwhile, Guo et al. (2022) explored how to bridge the gaps between BIM education and practice from a competency-based education perspective. In contrast, Ding et al. (2015) explore the key factors that affect practitioners' adoption of BIM. Their study showed that BIM motivation, BIM technical barriers, and BIM ability are the significant factors that affect architects' adoption of BIM technology. Howard et al. (2017) use the Unified Theory of Acceptance and Use of Technology (UTAUT) model to understand the individuals' perceptions towards working with BIM. The results reveal that performance expectancy does not directly affect behavioral intention of personal participation in BIM, signifying that BIM is perceived as an unrewarded addition to existing work processes.

Reviewing the existing literature, it is found that there is a relative lack of research on the learning behavior of BIM technology from the perspective of individual students. In recent years, exploring BIM education from students' perspectives has been encouraged. Agirbas (2020) pointed out that students' positive attitude towards using BIM software may improve their learning efficiency of professional architectural knowledge. In addition, through the evaluation of BIM learning achievement in China's Sixth National BIM Graduation Design Innovation Competition of Colleges and Universities, Ao et al. (2021) found seven public factors that affect BIM learning performance: (1) ability of the instructor, (2) school (college) atmosphere, (3) teamwork, (4) individual ability, (5) understanding of BIM industry applications, (6) social environment incentives, and (7) achievement demand. Furthermore, referring to the body of work that exists, this research combines the theory of UTAUT to analyze the influencing factors of college students' BIM learning intention and behavior and their influencing relationships from the perspective of individual learners, aiming at exploring and developing the path of BIM education.

2.2 Literature Review

2.2.1 Research Status of BIM Technology

In the past few decades, the construction industry has become increasingly interested in using Building Information Modeling (BIM) technologies (Jin et al., 2017). BIM is meant for the process of creating and managing information, which enables AEC professionals to effectively plan, design, construct, and manage the whole construction project's life cycle (Autodesk, 2021). The whole life cycle phases are indispensable in the application of BIM, and the lack of one aspect will considerably reduce the effectiveness of BIM use (Doherty, 2009). Therefore, BIM education should not be limited to modeling; students must probe into the BIM workflow (Tsai et al., 2019). Moreover, most of the research on BIM focuses on its application and implementation (Yalcinkaya & Singh, 2015), ignoring the perceived influence at the personal level (Howard et al., 2017).

2.2.2 Research Status of BIM Education

Currently, more and more attention is paid to BIM education (Moreno et al., 2019; Wang et al., 2020a). Chegu Badrinath et al. (2016) believe that BIM-educated students can visualize and integrate the project data well, which aligns with the current enterprise talent demand. The research and application of BIM are extensive, but some countries still lack BIM promotion. For instance, a study by Kolarić et al. (2019) showed that the application of BIM in Croatia and Slovakia has been deficient. Moreover, it is also believed that the improvement of BIM education will greatly impact improving BIM awareness and acceptance and will encourage those countries to integrate BIM technology into the AEC industry further. Scholars in many countries believe that BIM education needs to be improved through popularization of digitalization and improvement of BIM education that can positively affect contextualized learning and the improvement of BIM awareness and promote the development of BIM technology (Keung et al., 2022; Kolarić et al., 2019; Olowa et al., 2022; Sampaio, 2021; Sanchez et al., 2022; Sotelino et al., 2020). The overwhelming development trend of BIM technology makes colleges and universities pay more attention to cultivating BIM technical talents (Sun & Cao, 2021). BIM-into-extended reality (Virtual

Reality, Augmented Reality, Mixed Reality) has laid an excellent theoretical foundation for building a complete BIM education system in colleges and universities (Alizadehsalehi et al., 2019, 2020; 2021; Alizadehsalehi & Yitmen, 2021; Kang, 2020). However, the training scale of BIM technical talents is far from meeting the needs of industrial development, and BIM education cannot completely support the development of AEC industry (Jia, 2021; Wang, 2018). Although many colleges and universities have incorporated BIM-related courses into the curriculum (Brokbals & Cadez, 2017; Sacks & Barak, 2010; Wang et al., 2020b), Shi (2021) considers that the current BIM education cannot felicitously link BIM with other architectural fields in the curriculum. At the same time, people's shortage of knowledge of new technology leads to negative attitudes, and the consequences may be far-reaching, which will have a massive impact on technology adoption and industry development (Brown et al., 2017). Colleges and universities should realize that students' perception of BIM technology plays a crucial role in cultivating qualified BIM technical talents.

2.2.3 Unified Theory of Acceptance and Use of Technology Model

With the rapid development of information technology, people's intention to accept and use new technologies has always been a concern for researchers, who have put forward many theoretical models to predict people's intentions and behaviors. For example, Fishbein and Ajzen (1975) proposed the Theory of Reasoned Action (TRA), claiming that an individual's attitude towards behavior and subjective norms can influence behavior intention, thus further affecting the actual behavior. When theoretical models are applied to various fields, researchers usually choose suitable structures from many models, mainly ignoring the influence of the unmatched part of the model (Venkatesh et al., 2003). Therefore, Venkatesh et al. (2003) put forward the Unified Theory of Acceptance and Use of Technology (UTAUT) model, which integrates eight models, namely the Theory of Reasoned Action (TRA), Technology Acceptance Model (TAM), Motivational Model (MM), Theory of Planned Behavior (TPB), Combined TAM, TPB (C-TAM-TPB), Model of PC Utilization (MPCU), Innovation Diffusion Theory (IDT), and Social Cognitive Theory (SCT). Venkatesh tested all the above models and integrated them to form a more refined model, UTAUT, which was widely used by later researchers (Ameri et al., 2019; Andrews et al., 2021; Hossain et al., 2019; Li & Zhao, 2021; Salahshour Rad et al., 2019; VanDerSchaaf et al., 2021).

UTAUT identifies four key factors: performance expectancy, effort expectancy, social influence, and facilitating conditions. These four key factors jointly influence the users' behavioral intention and their behavior.

The behavioral intention is affected by the following variables:

1. Performance expectancy: The degree to which individuals feel that using this technology is helpful to their work or study.
2. Effort expectancy: Individuals feel the degree of effort required to use this technology.

3. Social influence: The degree to which individuals feel influenced by the surrounding groups.
4. The user behavior is affected by the following variables:
   (a) Behavioral intention: Individuals' judgment of the subjective possibility of using or learning this technology.
   (b) Facilitating conditions: The degree to which individuals feel supported by the organization when using or learning this technology.

The UTAUT model is widely used in the field of education to study students' acceptance of a specific technology or system. For example, Mohan et al. (2020) and Altalhi (2020) used the UTAUT model to analyze the application status of MOOCs in higher education, and the relationship between behavioral intention and behavior is deeply discussed. Chatterjee and Bhattacharjee (2020) applied the UTAUT model to explore the application of Artificial Intelligence (AI) in higher education. Alasmari and Zhang (2019) investigated college students' acceptance of mobile learning technology. They discovered that such variables as learning expectancy, effort expectancy, social influence, and characteristics of mobile learning are essential predictors of students' acceptance of mobile learning technology. At the same time, Sidik and Syafar (2020) have drawn a similar conclusion from the research on the factors affecting students' intention to use mobile learning. The results indicated that performance expectancy, effort expectancy, external influence, and service quality positively influence study intention. Abbad (2021) applied the UTAUT model to investigate the students' use of Moodle, an e-learning system. The findings revealed that performance expectancy and effort expectancy could influence students' behavioral intention to use Moodle (an E-learning system), and behavioral intention and facilitating conditions can directly influence their user behavior. Noble et al. (2022) employed the UTAUT model to show that performance expectancy, effort expectancy, and social influence can significantly affect the use intention of Virtual Reality (VR) learning. As shown, the UTAUT model has been widely used in education research. Therefore, this research applies the mature UTAUT model to explore the influential factors and correlation between college students' intention to learn BIM technology and their user behaviors, aiming to provide theoretical support for further BIM technology education.

2.3 Materials and Methods

2.3.1 Model Construction and Research Hypothesis

This research draws lessons from the existing UTAUT theoretical research results and combines the characteristics of college students, higher education, and BIM technology learning to construct a model of influencing factors of college students' BIM technology learning intention and behavior. This research model retains the four key factors in the classic UTAUT model: performance expectancy, effort expectancy, social influence and facilitating conditions. "Attitude" is also considered a very strong factor affecting behavioral intention (Botero et al., 2018; Davis et al., 1989), so this research considers "attitude" a critical variable in the model. In addition, Dwivedi et al. (2017) also mention that most studies using UTAUT often gave up the use of regulatory variables. Arif et al. (2018) did not observe the moderating effect of age, gender, and experience in the study of influencing factors on students' use of distance education based on network services. At the same time, because BIM technology is a compulsory course for AEC majors in most Chinese universities, age, gender, experience, and voluntariness were excluded from the regulated variables. Therefore, the theoretical framework model proposed in this research is shown in Fig. 2.1.

By summarizing the existing research results, this study puts forward the following research hypotheses, as shown in Table 2.1.

2.3.2 *Questionnaire Design*

Based on the scale recorded in the existing literature, combined with the BIM learning situation of Chinese AEC majors, this study's designed questionnaire mainly includes two parts: the first part records the basic information of AEC majors in universities in China, including the interviewees' gender, age, grade, major, BIM knowledge, and access to BIM courses, and so forth. The second part measures students' intention and behavior to learn BIM and adopts a Likert five-component scale to measure the degree of each observed variable with 1–5 indicating "totally disagree", "basically disagree", "average", "basically agree", and "totally agree" respectively. The measurement scale of the questionnaire mainly comes from the UTAUT scale proposed by Venkatesh et al. (2003) and the perceived usefulness scale by Davis et al. (1989). Considering the BIM learning context, the learning behavior scale of this study contains 21 items. In addition, a polygraph question is set in the questionnaire: "Please choose the option of completely disagreeing with this

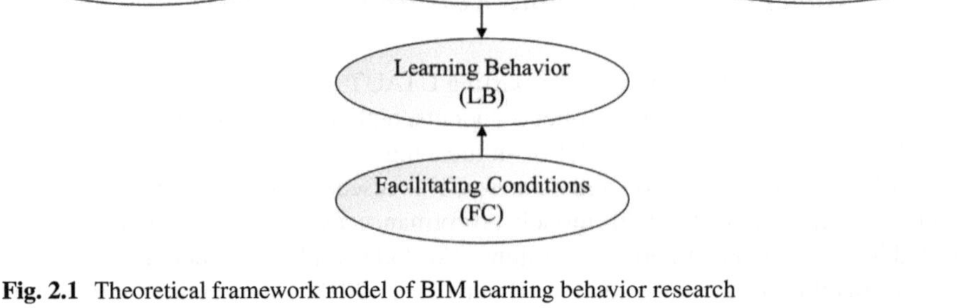

Fig. 2.1 Theoretical framework model of BIM learning behavior research

Table 2.1 Research hypotheses

Hypothesis number	Research hypothesis	References
H1	College students' attitude toward BIM technology will positively affect their learning intention	Andrews et al. (2021)
H2	College students' performance expectancy of BIM technology will positively affect their learning intention	Chang (2013); Li and Zhao (2021); Venkatesh et al. (2003)
H3	College students' efforts expectancy of BIM technology will positively affect their learning intention	Chang (2013); Li and Zhao (2021); Venkatesh et al. (2003)
H4	College students' social influence on BIM technology will positively affect their learning intention	Chang (2013); Li and Zhao (2021); Venkatesh et al. (2003)
H5	College students' facilitating conditions of BIM technology will positively affect their learning behavior	Abbad (2021); Venkatesh et al. (2003)
H6	College students' behavioral intention toward BIM technology will positively affect their learning behavior	Abbad (2021)

question". If the respondent chooses other options, it is considered that there are some cases, such as choosing directly without reading the questions, and the questionnaire is regarded as invalid. Please see Appendix A for the detailed questionnaire.

2.3.3 Sampling and Date Collection

This study conducted a questionnaire survey in March 2022 using random sampling. The process was divided into three steps: choosing the investigation area, schools, and students.

1. Sampling investigation area: In order to make the research data representative and convincing, the sampled universities cover seven geographical regions in China, including northeast China, north China, east China, south China, central China, northwest China, and southwest China.
2. Sampling colleges and universities: Since BIM technology mainly involves AEC-related majors, universities with civil engineering, engineering management, and other construction majors were mainly selected in the school selection process. According to the types of undergraduate colleges in China, at least one university of four types (research-oriented, applied research-oriented, application-oriented, and private research-oriented) was randomly selected from seven geographical regions to ensure that the sample universities cover the seven regions and four types. We also started point-to-point contact with randomly

selected universities to ensure their intention to cooperate. Finally, 35 sample universities were determined, including 8 research-oriented universities, 10 applied research-oriented universities, 8 application-oriented universities, and 9 private research-oriented universities. Sampling information of colleges and universities is shown in Table 2.2.

Table 2.2 Sample school selection results

Areas	Research-oriented university	Applied research-oriented university	Application-oriented University	Private application-oriented university
Central China	Wuhan University	China Three Gorges University	Zhengzhou University of Aeronautics	College of Science and Technology of China Three Gorges University
		Hunan University of Science and Technology		
		Changsha University of Science and Technology		
East China	Hefei University of Technology	Shandong Jianzhu University	Suzhou University of Science and Technology	Tianping College of Suzhou University of Science and Technology
				Anhui Xinhua University
North China	Beijing Jiaotong University	Shijiazhuang Tiedao University	Inner Mongolia University of Science and Technology	Shanxi Technology and Business University
	Central University of Finance and Economics			
Northeast China	Northeast Forestry University	Shenyang Jianzhu University	Dalian Ocean University	East University of Heilongjiang
			Liaoning University of Technology	Jilin University of Architecture and Technology
Northwest China	Chang'an University	Xi'an University of Architecture and Technology	HeXi University	Ningxia Institute of Science and Technology
Southwest China	Chongqing University	Southwest University Of Science And Technology	Leshan Normal University	Southwest Jiaotong University Hope College
		Kunming University of Science and Technology		
South China	South China University of Technology	Hainan University	Wuyi University	University of Sanya

3. Sampling students: In this study, online questionnaires were distributed among the students via the teachers from the selected colleges. The research intended to choose at least 30 valid samples in each college, and it was estimated that 1050 valid total samples would be selected.

Data collection mainly went through two stages: the first stage was the pre-investigation stage, which aimed to get feedback to improve the questionnaire quality. This stage started on March 20 and lasted until March 26, 2022. The second stage was the formal investigation stage, which lasted 6 days, from March 27 to April 3, 2022. The research team contacted AEC-related professional teachers from the selected schools to request they issue online questionnaires. The research team kept tracking and feeding back the questionnaire results, and finally, 1699 questionnaires were collected. A total of 609 invalid questionnaires were eliminated through polygraph questions, and 1090 valid questionnaires were finally recovered with an effectiveness rate of 64.16%.

Among the 1090 valid questionnaires, 455 were male, accounting for 41.7% of the total and 635 were women, accounting for 58.3% of the total population. The interviewees' age is mainly between 19 and 22 years old, which is also in line with the age group of undergraduate students. In terms of grades, sophomores account for the most, accounting for 33%; juniors for the next, accounting for 27%; freshmen and seniors for the least, each accounting for 20%. First-year students have just entered the university, so they do not know much about BIM technology and other professional technologies. Hence, their intention to fill out the questionnaire was not firm. Moreover, most seniors have entered internship positions, and the samples collected were few. Table 2.3 represents the interviewees' statistics.

2.4 Results and Findings

2.4.1 Reliability and Validity Test

This study used Cronbach's Alpha coefficient for the reliability test (Bagozzi & Yi, 1988). Generally speaking, Cronbach's Alpha coefficient should be greater than 0.7 (Schmitt, 1996). The Cronbach's Alpha coefficient of this research data was 0.913, so the reliability of this scale is quite good. In addition, the KMO value of the scale was 0.885, which is greater than 0.7, and the p-value in Bartlett Sphericity test is less than 0.05, which is significant. This shows that this scale passed the KMO and Bartlett Sphericity tests and is suitable for factor analysis.

AMOS24.0 software was used to establish the measurement model in this study, and the maximum likelihood estimation method was used for confirmatory factor analysis. The analysis results showed that the convergent validity passed the Average Variance Extracted (AVE) test (AVE > 0.5), and the structure had good convergence (Joe F. Hair et al., 2012). In addition, when the composite reliability (CR) is greater than 0.7, the scale has good convergence validity (Fornell & Larcker, 1981). The

Table 2.3 Descriptive statistics

Topic item	Option	Count Heads	Percentage of the total number of people
Gender	Male	455	42%
	Female	635	58%
Age	19 years or younger	58	5%
	19–21 years old	749	69%
	22–24 years old	280	26%
	Over 24 years old	3	0.3%
Grade	Freshman	214	20%
	Sophomore	357	33%
	Junior	299	27%
	Senior	220	20%
Specialized subject	Engineering management	553	50.7%
	Construction costs	247	22.7%
	Civil engineering	94	8.6%
	Architecture	60	5.5%
How do you know the BIM industry?	Understand	255	24%
	Common	232	21%
	Incomprehension	603	55%
Whether you have participated in BIM competitions or not?	Yes	188	17%
	No	902	83%
Whether the school offer BIM courses or not?	Yes	813	75%
	No	277	25%

test results of convergent validity of each scale are shown in Table 2.4. In Table 2.5, SQRT (AVE) is larger than the correlation coefficients on the horizontal and vertical lines of the table; hence, it shows that the scale has good discriminant validity. To sum up, the reliability and validity of this research scale were tested, and it can be further analyzed.

2.4.2 *Goodness of Fit*

It can be concluded from the above table that the theoretical model is fitted in this study, and the Chi-square freedom ratio was 4.541, less than 5, which is within the acceptable range. Moreover, RMSEA was 0.057, less than 0.08, and the fitting degree was good. GFI, AGFI, NFI, TLI, and CFI were all greater than 0.9, so the structural equation model in this study has a good fitting degree. Table 2.6 shows the specific indicators for fitting the research structure model.

Table 2.4 Results table of confirmatory factor analysis

Variable	Topic item	Factor loading	CR	AVE
performance expectancy	PE1	0.808	0.911	0.773
	PE2	0.946		
	PE3	0.879		
effort expectancy	EE1	0.754	0.873	0.698
	EE2	0.881		
	EE3	0.865		
social influence	SI1	0.881	0.860	0.675
	SI2	0.888		
	SI3	0.678		
facilitating conditions	FC1	0.74	0.850	0.586
	FC2	0.823		
	FC3	0.749		
	FC4	0.747		
learning attitude	LA1	0.835	0.916	0.786
	LA2	0.883		
	LA3	0.938		
learning intention	LI1	0.86	0.866	0.683
	LI2	0.833		
	LI3	0.785		
learning behavior	LB1	0.897	0.891	0.803
	LB2	0.895		

2.4.3 *Path Analysis*

The fitting results of the model are shown in Fig. 2.2.

It can be seen from Fig. 2.2 that the significance of all paths was less than 0.05, except that efforts expectancy was greater than 0.05, and the standardized regression coefficients were all positive. Among them, the intention to learn BIM was influenced by performance expectancy, social influence, and learning attitude, and the order of influence was learning attitude (0.675) > social influence (0.144) > performance expectancy (0.101). Study behavior was influenced by facilitating conditions (0.212) and behavioral intention (0.203).

In order to explore the influence of exogenous variables on BIM technology learning behavior, this study used "indirect", "direct", and "total effects" in the "output" select interface of AMOS24.0 to further analyze the direct, indirect, and total effects of the theoretical model path, and the results are shown in Table 2.7. Among them, learning attitude, social influence, and performance expectancy significantly indirectly affect learning behavior through learning intention. Compared with social influence and performance expectancy, the indirect effect of learning attitude on learning behavior is more substantial.

Table 2.5 Table of distinctions

	Performance expectancy	Effort expectancy	Social influence	Facilitating conditions	Learning attitude	Learning intention	Learning behavior
performance expectancy	0.879						
effort expectancy	0.276 ***	0.835					
social influence	0.472 ***	0.368 ***	0.822				
facilitating conditions	0.507 ***	0.404 ***	0.537 ***	0.766			
learning attitude	0.389 ***	0.357 ***	0.420 ***	0.479 ***	0.887		
learning intention	0.424 ***	0.280 ***	0.464 ***	0.444 ***	0.763 ***	0.826	
learning behavior	0.091 **	0.380 ***	0.175 ***	0.306 ***	0.270 ***	0.293 ***	0.896

Note: ***, $p < 0.001$; **, $p < 0.01$; Diagonal line is SQRT (AVE)

Table 2.6 Structural model fitting index

Index	CMIN/DF	GFI	AGFI	RMSEA	NFI	TLI	CFI
Text value	4.541	0.934	0.911	0.057	0.948	0.950	0.959
Recommended value	<5	>0.9	>0.9	<0.08	>0.9	>0.9	>0.9

2.5 Discussion

Based on UTAUT theory, this study explores the factors influencing students' learning behavior toward BIM technology in AEC-related majors. However, it is worth noting that effort expectancy has no significant effect on learning intention.

Firstly, through the verification of BIM technology learning intention and behavior model, we found that learning attitude ($\beta = 0.675$, $p < 0.001$) has the greatest predictive effect on college students' BIM technology learning intention, while learning attitude ($\beta = 0.137$, $p < 0.001$) has a significant indirect impact on learning behavior through learning intention. In this study, among the three observed variables that constitute attitudes, LA2 (I am interested in BIM technology) and LA3 (It is fun to learn BIM technology) have the highest factor loads. It can be seen that when college students are more interested in BIM technology, their intention to learn is higher. This finding is similar to the conclusion drawn by Mohan et al. (2020), manifesting that students are more willing to learn when they think MOOCs are more attractive.

Furthermore, this study also verified that performance expectancy ($\beta = 0.101$, $p < 0.001$) and social influence ($\beta = 0.144$, $p < 0.002$) have a significant direct influence on college students' intention to learn BIM technology. In addition,

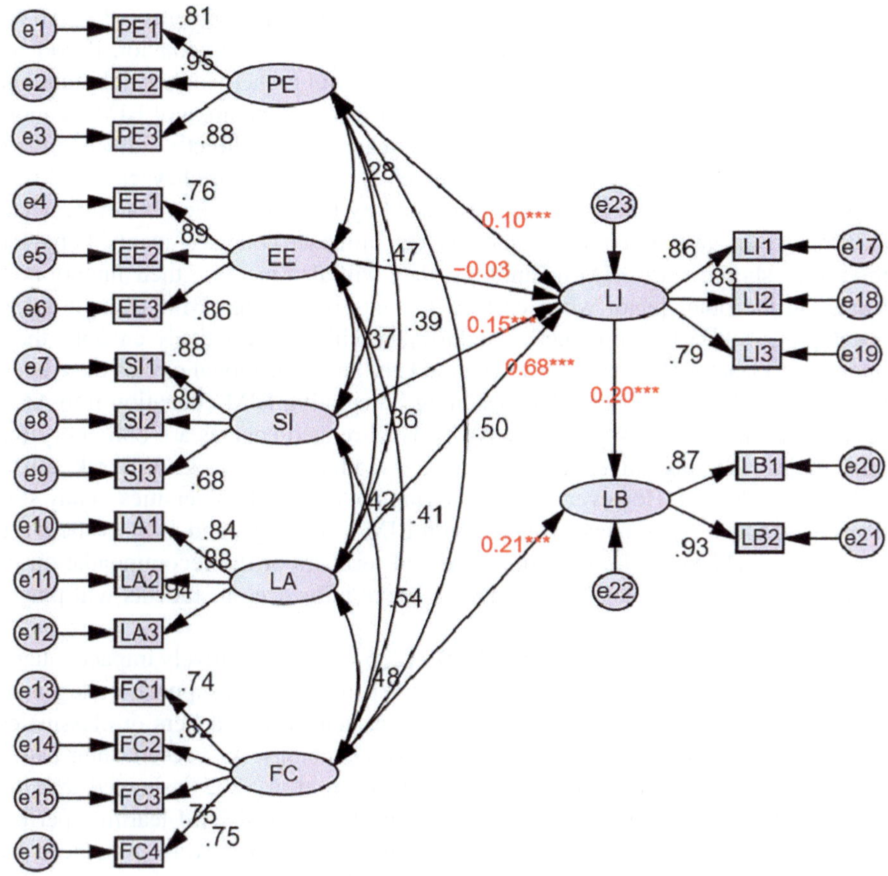

Fig. 2.2 Model analysis diagram of BIM learning behavior research. Note: ** means less than 0.001 < p < 0.005, and *** means less than p < 0.001

Table 2.7 Total effect, direct effect, and indirect effect

Variable	Effect Type	AT	SI	EE	PE	FC	LI
LI	Total	0.675 ***	0.144 **	−0.035	0.101 ***	–	–
	Direct	0.675 ***	0.144 **	−0.035	0.101 ***	–	–
	Indirect	–	–	–	–	–	–
LB	Total	0.137 ***	0.029 ***	−0.007	0.02 ***	0.212 ***	0.203 ***
	Direct	–	–	–	–	0.212 ***	0.203 ***
	Indirect	0.137 ***	0.029 ***	−0.007	0.02 ***	–	–

Note: ** means less than $0.001 < p < 0.005$, and *** means less than $p < 0.001$

performance expectancy ($\beta = 0.02$, $p < 0.001$) and social influence ($\beta = 0.029$, $p < 0.002$) have a considerable indirect influence on learning behavior. If college students think that learning BIM technology is helpful to their studies, job hunting, and career development, they will have a higher learning intention of BIM technology. The higher learning intention will significantly and indirectly influence their BIM technology learning behavior. Research by Choukas-Bradley et al. (2015) shows that young people are easily influenced by their peers. For example, when classmates, teachers, and other people who can influence college students' behavior think they should learn BIM technology, they will also increase their intention to learn. On the other hand, when college students realize that BIM technology is developing well in the construction industry, it will also effectively enhance their intention to learn BIM technology. The industry and educational circles have been working together to improve BIM teaching, integrating BIM education into AEC courses in universities (Wu & Issa, 2014). BIM-related courses are obligatory for AEC majors in many universities (Brokbals & Cadez, 2017; Sacks & Barak, 2010; Wang et al., 2020b). However, before entering colleges and universities, many students have never experienced working with BIM technology, and most of them are exposed to BIM technology at the request of teachers or the recommendation of senior seniors. Under such conditions, the social influence on students will play a great role.

Moreover, facilitating conditions ($\beta = 0.212$, $p < 0.001$) positively impact college students' BIM technology learning behavior. When the school supports college students in learning BIM technology, they can get help from teachers or classmates when they encounter difficulties learning BIM technology. As a result, their learning can be improved. For example, research by Ao et al. (2021) showed that the teacher's guidance significantly positively impacts students' BIM learning performance. In addition, students' learning behavior can be positively influenced by facilitating their access to the required resources from the school library or online resources while learning BIM technology.

However, unlike the conclusion of Venkatesh et al. (2003), in this study, effort expectation has no significant influence on college students' intention to learn BIM technology ($\beta = -0.035$, $p > 0.05$). This result is surprising because it is inconsistent with most previous research conclusions. For example, the research of Li and Zhao (2021) reflects that effort expectancy is expected to have a significant positive impact on the intention of continuing to use MOOC. VanDerSchaaf et al. (2021) research also shows that effort expectancy and social influence are the key influences on college students' intention to use information technology to obtain university services. However, the situation of this study is not without precedent. For instance, in the study of influencing factors of doctors' adoption of electronic health records by Hossain et al. (2019), it was concluded that efforts are expected to have no significant impact on doctors' adoption of electronic health records. Andrews et al. (2021) found that the effort is expected to have no significant impact on librarians' adoption of AI and related technologies. This work (Salahshour Rad et al., 2019) suffers the same limitations as social networking sites, and effort expectancy has no significant influence on behavior intention. In the study of Indian graduate students' intention and obstacles using MOOCs by Mohan et al. (2020), relevant conclusions have also been drawn. Efforts expectancy, social influence, and facilitating conditions have no statistically significant influence on the intention to use MOOCs. In the research of pharmaceutical students' acceptance of LabSafety based on mobile devices by Ameri et al. (2019), it was also found that effort expectancy has no significant influence on their intention to use. Further analysis shows that there might be two reasons for the insignificant influence of effort expectation on learning intention in this study: First, college students are generally younger (ranging from 19 to 25 years old). Therefore, they can easily accept the new technology, and it is less cumbersome to learn BIM technology since potential hardship in learning new technologies can not affect their intention to learn. Second, only basic software operations are being taught in many schools. Many complex functions have yet to be explored by students. Therefore, it is not clear what efforts are needed to learn BIM technology, and this issue has been mentioned by many students in questionnaire interviews. As shown in the scores of the performance expectancy items, the scores are mainly distributed between "general" and "basic agreement", which is closer to "general", and it can be seen that most students do not know much about BIM technology's performance expectancy or perceived ease of use. In other words, students do not know much about the difficulty of BIM technology. They are only informed about BIM technology's capabilities to obtain better job opportunities, but they do not know exactly how and what to learn. Therefore, some college students' understanding of BIM technology is still insufficient and there is no significant statistical relationship between their expectation and intention to learn BIM technology.

2.6 Conclusion and Implications

Through sorting out and analyzing a large number of related literature works, this study uses UTAUT to construct the model of influencing factors of BIM technology learning intention and behavior from the perspective of students' perceptions. A random sample questionnaire survey was conducted among AEC majors in four types of universities (research-oriented, applied research-oriented, application-oriented, and private research-oriented) in seven regions of China (northeast, north, east, south, central, northwest, and southwest), and the influencing factors and their relationships were explored. The results show that, in addition to effort expectancy, performance expectancy, social influence, and learning attitude have significant direct effects on learning intention and have significant indirect effects on learning behavior through learning intention; facilitating conditions and learning intention have significant direct effects on learning behavior. This study once again verified the effectiveness of UTAUT in the field of BIM technology learning and confirmed the significant influence of attitude on BIM technology learning intention and behavior.

2.6.1 Practical Implications

Based on the research conclusion, this study puts forward the following suggestions for further improving the BIM technology education for college students:

1. Given the significant positive influence of learning attitude on learning intention and behavior, the learning objectives of undergraduate students need to be clarified to improve the learning attitude towards BIM technology. Therefore, to make students' learning objectives clear, the curriculum arrangement in colleges and universities should include basic knowledge of BIM and career planning. In this way, first-year students can overcome their fear of ignorance of BIM industry. Moreover, a well-established BIM curriculum may assist students in realizing the importance of BIM technology to the future of AEC industry.
2. Because of the significant positive influence of facilitating conditions on learning behavior, the BIM technology training for teachers should be strengthened to improve BIM technology operation and teaching ability. The more the teachers are capable of teaching BIM technology, the more support they can provide for the students. Thus, the improvement of teachers' professional skills will be beneficial to BIM technology education. At present although, college teachers are primarily the young generation with scientific solid research abilities, their teaching ability is generally not enough due to the limitation of working time and experience (Ting-ting et al., 2022). Therefore, college teachers must carry out teaching skills training to improve their professional quality.
3. Given the significant positive influence of facilitating conditions on learning behavior, resources and support for learning BIM technology ought to be enhanced. The leading exporters of BIM talents, colleges, and universities should provide comprehensive support for the application of BIM technology from the aspects of system, human resources, funds, and policies (Jin, 2019). In addition to ensuring the quality of BIM courses taught in schools, colleges, and universities, those organizations should encourage and support students to participate in various BIM competitions to improve undergraduate students' BIM application ability. This suggestion may lead to continuously training qualified AEC graduates for the industry.
4. In view of the significant positive influence of social influence on learning intention and behavior, the promotion of BIM industry and the creation of BIM learning and application atmosphere are vital. The higher the industry adoption of BIM technology, the stronger the willingness of college students to learn. Hence, the government should provide policy and economic support to enterprises, encourage the construction industry to apply BIM technology, and actively promote BIM technology, creating a solid learning atmosphere for BIM technology educators and learners.

2.6.2 Future Work

This study adopts the UTAUT model, which has a very high degree of explanation for user intention and behavior. However, the model does not consider individual students' learning abilities and methods. Future directions of work focus on adding constructivist learning theory and considering learning self-efficacy on this research basis. In addition, considering that too many questions in the online questionnaire may seem boring for respondents, the number of questionnaire items is relatively small. However, fewer items in each latitude may make deeper content impossible to explore. In the following research, each dimension can be further subdivided in detail, and additional items can be added to make the measurement results more accurate.

References

Abbad, M. M. M. (2021). Using the UTAUT model to understand students' usage of e-learning systems in developing countries. *Education and Information Technologies, 26*(6), 7205–7224.

Agirbas, A. (2020). Teaching construction sciences with the integration of BIM to undergraduate architecture students. *Frontiers of Architectural Research, 9*(4), 940–950.

Alasmari, T., & Zhang, K. (2019). Mobile learning technology acceptance in Saudi Arabian higher education: An extended framework and a mixed-method study. *Education and Information Technologies, 24*(3), 2127–2144.

Alizadehsalehi, S., & Yitmen, I. (2021). Digital twin-based progress monitoring management model through reality capture to extended reality technologies (DRX). *Smart and Sustainable Built Environment, 12*, 200–236.

Alizadehsalehi, S., Hadavi, A., & Huang, J. (2019). Virtual reality for design and construction education environment. *Journal of Construction Engineering and Management.*

Alizadehsalehi, S., Hadavi, A., & Huang, J. C. (2020). From BIM to extended reality in AEC industry. *Automation in Construction, 116*, 103254.

Alizadehsalehi, S., Hadavi, A., & Huang, J. C. (2021). Assessment of AEC students' performance using BIM-into-VR. *Applied Sciences, 11*(7).

Altalhi, M. (2020). Toward a model for acceptance of MOOCs in higher education: The modified UTAUT model for Saudi Arabia. *Education and Information Technologies, 26*(2), 1589–1605.

Ameri, A., Khajouei, R., Ameri, A., & Jahani, Y. (2019). Acceptance of a mobile-based educational application (LabSafety) by pharmacy students: An application of the UTAUT2 model. *Education and Information Technologies, 25*(1), 419–435.

Andrews, J. E., Ward, H., & Yoon, J. (2021). UTAUT as a model for understanding intention to adopt AI and related technologies among librarians. *The Journal of Academic Librarianship, 47*(6), 102437.

Ao, Y., Liu, Y., Tan, L., Tan, L., Zhang, M., Feng, Q., Zhong, J., Wang, Y., Zhao, L., & Martek, I. (2021). Factors driving BIM learning performance: Research on China's sixth national BIM graduation design innovation competition of colleges and universities. *Buildings, 11*(12).

Arif, M., Ameen, K., & Rafiq, M. (2018). Factors affecting student use of web-based services. *The Electronic Library, 36*(3), 518–534.

Autodesk (2021). What is BIM, building information modeling.

Bagozzi, R. R., & Yi, Y. (1988). On the evaluation of structural equation models. *Journal of the Academy of Marketing Science, 16*(1), 74–94.

Botero, G. G., Questier, F., Cincinnato, S., He, T., & Zhu, C. (2018). Acceptance and usage of mobile assisted language learning by higher education students. *Journal Of Computing In Higher Education, 30*(3), 426–451.

Brokbals, S., & Cadez, I. (2017). Academic teaching of BIM – Development – Status quo – Demand for action. *Bautechnik, 94*(12), 851–856.

Brown, S. A., Massey, A. P., Montoya-weiss, M. M., & Burkman, J. R. (2017). Do I really have to? User acceptance of mandated technology. *European Journal of Information Systems, 11*(4), 283–295.

Chang, C.-C. (2013). Library mobile applications in university libraries. *Library Hi Tech, 31*(3), 478–492.

Chatterjee, S., & Bhattacharjee, K. K. (2020). Adoption of artificial intelligence in higher education: A quantitative analysis using structural equation modelling. *Education and Information Technologies, 25*(5), 3443–3463.

Chegu Badrinath, A., Chang, Y. T., & Hsieh, S. H. (2016). A review of tertiary BIM education for advanced engineering communication with visualization. *Visualization in Engineering, 4*(1).

Choukas-Bradley, S., Giletta, M., Cohen, G. L., & Prinstein, M. J. (2015). Peer influence, peer status, and prosocial behavior: An experimental investigation of peer socialization of adolescents' intentions to volunteer. *Journal of Youth and Adolescence, 44*(12), 2197–2210.

Davis, F. D., Bagozzi, R. P., & Warshaw, P. R. (1989). User acceptance of computer technology: A comparison of two theoretical models. *Management Science, 35*(8), 982–1003.

Ding, Z., Zuo, J., Wu, J., & Wang, J. Y. (2015). Key factors for the BIM adoption by architects: A China study. *Engineering, Construction and Architectural Management, 22*(6), 732–748.

Doherty, G. E. (2009). *How BIM and integrated practice may change architectural, engineering, and construction education* (Master). University of Wyoming.

Dwivedi, Y. K., Rana, N. P., Jeyaraj, A., Clement, M., & Williams, M. D. (2017). Re-examining the unified theory of acceptance and use of technology (UTAUT): Towards a revised theoretical model. *Information Systems Frontiers, 21*(3), 719–734.

Eadie, R., Browne, M., Odeyinka, H., McKeown, C., & McNiff, S. (2013). BIM implementation throughout the UK construction project lifecycle: An analysis. *Automation in Construction, 36*, 145–151.

Ebrahimi, P. R., Alizadehsalehi, S., & Mosaberpanah, M. A. (2019). Interactions of sustainability and BIM in support of existing buildings. *International Conference on Sustainable Infrastructure, 2019*, 216–225.

Fishbein, M., & Ajzen, I. (1975). *Belief, attitude, intention, and behavior: An introduction to theory and research*. Addison-Wesley.

Fornell, C., & Larcker, D. F. (1981). Evaluating structural equation models with unobservable variables and measurement error. *Journal of Marketing Research, 18*(1), 39.

Guo, B. H. W., Gonzalez, V. A., Puolitaival, T., Enegbuma, W., & Zou, Y. (2022). Bridging the gap between building information modelling education and practice: A competency-based education perspective. *International Journal of Construction Management*, 1–12.

Hair, J. F., Sarstedt, M., Ringle, C. M., & Mena, J. A. (2012). An assessment of the use of partial least squares structural equation modeling in marketing research. *Journal of the Academy of Marketing Science, 3*(40), 414–433.

Hossain, A., Quaresma, R., & Rahman, H. (2019). Investigating factors influencing the physicians' adoption of electronic health record (EHR) in healthcare system of Bangladesh: An empirical study. *International Journal of Information Management, 44*, 76–87.

Howard, R., Restrepo, L., & Chang, C.-Y. (2017). Addressing individual perceptions: An application of the unified theory of acceptance and use of technology to building information modelling. *International Journal of Project Management, 35*(2), 107–120.

Jia, X. (2021). Research and education status analysis of building information Technology in China. *Real Estate World, 09*, 29–33.

Jin, C. (2019). *Understanding the adoption and continuous use of flipped classroom by instructors and students from the perspective of teaching system* (Doctoral dissertation). Central China Normal University, Hubei Province.

Jin, R., Hancock, C. M., Tang, L., & Wanatowski, D. (2017). BIM investment, returns, and risks in China's AEC industries. *Journal of Construction Engineering and Management, 143*(12).

Kang, S. (2020). *Research on BIM education performance evaluation of engineering management major in colleges and universities* (Master of engineering). Yangzhou University, Jiangsu province.

Keung, C. C. W., Yiu, T. W., & Feng, Z. (2022). Building information modeling education for quantity surveyors in Hong Kong: Current states, education gaps, and challenges. *International Journal of Construction Education and Research, 19*, 1–17.

Kolarić, S., Mandičák, T., Vukomanović, M., & Mesároš, P. (2019). The influence of BIM education improvement on raising BIM awareness in Croatia and Slovakia. *Periodica Polytechnica Architecture, 50*(1), 21–29.

Li, Y., & Zhao, M. (2021). A study on the influencing factors of continued intention to use MOOCs: UTAUT model and CCC moderating effect. *Frontiers in Psychology, 12*.

Mohan, M. M., Upadhyaya, P., & Pillai, K. R. (2020). Intention and barriers to use MOOCs: An investigation among the post graduate students in India. *Education and Information Technologies, 25*(6), 5017–5031.

Moreno, C., Olbina, S., & Issa, R. R. (2019). BIM use by architecture, engineering, and construction (AEC) industry in educational facility projects. *Advances in Civil Engineering, 2019*, 1–19.

Noble, S. M., Saville, J. D., & Foster, L. L. (2022). VR as a choice: What drives learners' technology acceptance? *International Journal of Educational Technology in Higher Education, 19*(1).

Olowa, T., Witt, E., Morganti, C., Teittinen, T., & Lill, I. (2022). Defining a BIM-enabled learning environment – An adaptive structuration theory perspective. *Buildings, 12*(3).

Sacks, R., & Barak, R. (2010). Teaching building information modeling as an integral part of freshman year civil engineering education. *Journal of Professional Issues in Engineering Education and Practice, 136*(1), 30–38.

Salahshour Rad, M., Nilashi, M., Mohamed Dahlan, H., & Ibrahim, O. (2019). Academic researchers' behavioural intention to use academic social networking sites: A case of Malaysian research universities. *Information Development, 35*(2), 245–261.

Sampaio, A. Z. (2021). BIM education required in construction industry. In *Sustainability and automation in smart constructions* (pp. 3–9).

Sanchez, B., Ballinas, R., & Rodriguez-Paz, M. (2022). BIM and game engines for engineering online learning. In *Proceedings of the 2022 IEEE Global Engineering Education Conference (EDUCON)* (pp. 1467–1472). IEEE.

Schmitt, N. (1996). Uses and abuses of coefficient alpha. *Psychological Assessment, 4*(8), 350–353.

Shi, S. (2021). Research on BIM course system of civil engineering under the new engineering. *Sichuan University of Arts and Science Journal, 31*(2), 54–57.

Sidik, D., & Syafar, F. (2020). Exploring the factors influencing student's intention to use mobile learning in Indonesia higher education. *Education and Information Technologies, 25*(6), 4781–4796.

Sotelino, E. D., Natividade, V., & Travassos do Carmo, C. S. (2020). Teaching BIM and its impact on young professionals. *Journal of Civil Engineering Education, 146*(4), 05020005.

Sun, Y., & Cao, S. (2021). Factors of willingness to learn BIM technology in college students based on technology acceptance model. *Project Management Technology, 19*(3), 70–76.

Ting-ting, P., Ying, C., Bao-peng, X., & Jun, Z. (2022). Research on the influencing factors of young teachers' ability in colleges and universities' teaching – Take a university in Gansu Province as an example. *Journal of Hubei Normal University (Philosophy and Social Sciences), 42*(2), 112–117.

Tsai, M.-H., Chen, K.-L., & Chang, Y.-L. (2019). Development of a project-based online course for BIM learning. *Sustainability, 11*(20).

VanDerSchaaf, H. P., Daim, T. U., & Basoglu, N. A. (2021). Factors influencing student information technology adoption. *IEEE Transactions on Engineering Management*, 1–13.
Venkatesh, V., Morris, M. G., Davis, G. B., & Davis, F. D. (2003). User acceptance of information technology: Toward a unified view. *MIS Quarterly, 27*(3), 425–478.
Wang, S. (2018). The applied undergraduate civil engineering major is based on BIM practical teaching research of design competition mechanism. *Information Recording Materials, 19*(1), 140–141.
Wang, L., Huang, M., Zhang, X., Jin, R., & Yang, T. (2020a). Review of BIM adoption in the higher education of AEC disciplines. *Journal of Civil Engineering Education, 146*(3), 06020001.
Wang, L., Yan, X., Fan, B., Jin, R., Yang, T., & Kapogiannis, G. (2020b). Incorporating BIM in the final semester undergraduate project of construction management – A case study in Fuzhou university. *KSCE Journal of Civil Engineering, 24*(8), 2403–2418.
Wu, W., & Issa, R. R. A. (2014). BIM education and recruiting: Survey-based comparative analysis of issues, perceptions, and collaboration opportunities. *Journal of Professional Issues in Engineering Education and Practice, 140*(2).
Yalcinkaya, M., & Singh, V. (2015). Patterns and trends in building information modeling (BIM) research: A latent semantic analysis. *Automation in Construction, 59*, 68–80.
Yan, G., Wu, X., & Zhu, L. (2021). Exploration on BIM applied talents training mode of school-enterprise cooperation. *Housing and Real Estate, 3*, 251–252.

Chapter 3
Heterogeneity Analysis of the Driving Model for BIM Education Effectiveness

Abstract With the widespread adoption of Building Information Modeling (BIM) technology in the global construction industry, BIM education in higher education has become an integral part of construction management programs. Drawing upon Bandura's Social Cognitive Theory (SCT), this study develops a model to identify the influencing factors of BIM education effectiveness in construction management disciplines. The results indicate that the effectiveness of BIM education is shaped by multiple factors, including theoretical courses, practical activities, faculty resources, and institutional incentives. Moreover, significant differences were identified between industry professionals and current students regarding their perceptions of and demands for BIM education. Based on these findings, the study proposes strategies to optimize BIM education in Chinese universities, with particular emphasis on government–university–industry collaboration, aiming to enhance the practicality and effectiveness of BIM training and better align talent cultivation with industry needs.

Keywords BIM education · Construction management · Educational effectiveness · Structural equation modeling

3.1 Introduction

As a core technology of digital construction, Building Information Modeling (BIM) is profoundly reshaping the production logic of the architecture, engineering, and construction (AEC) industry. Since the Ministry of Housing and Urban-Rural Development (MOHURD) of China first emphasized the promotion of BIM in the 2011–2015 Construction Industry Informatization Development Outline in 2011, the strategic importance of this technology has continued to rise. It has since been

Y. Ao et al., *Building Information Modeling (BIM) Adoption*, Resilient and Friendly Living Environments, https://doi.org/10.1007/978-981-95-3993-2_3

repeatedly highlighted in the Opinions on Accelerating the Development of New-Type Construction Industrialization (2020) and the 14th Five-Year Development Plan for the Construction Industry, gradually forming policy continuity and broad industry consensus. Driven by policy guidance, industrial demand, and technological advancement, BIM has evolved from a simple modeling tool into a systematic platform supporting whole-process management, cross-disciplinary collaboration, and full life-cycle optimization. This evolution imposes new demands on higher education: within limited academic years, how can construction management programs cultivate professionals who not only master technical operations but also possess integrative thinking, teamwork competence, and strategic awareness? This question has become a shared concern of both academia and industry.

Although BIM technology has been widely adopted globally, a structural imbalance remains between talent supply and demand. Many enterprises report that graduates are often confined to software-level skills and lack the ability to apply BIM comprehensively in real projects. Meanwhile, university instructors frequently point to limited teaching resources and slowly updated curricula, which cannot keep pace with the rapidly evolving needs of industry. More critically, educators have an insufficient understanding of students' learning motivations, behavioral trajectories, and psychological mechanisms. Teaching design tends to be goal-oriented, but it fails to reveal students' internal drivers and external constraints during the learning process. At the same time, high-intensity, task-driven BIM competitions have emerged as an important bridge linking classrooms with industry. Yet, their educational value and mechanisms of influence remain underexplored through systematic empirical research. These issues suggest that course design and faculty input alone cannot fully explain or enhance BIM education outcomes; rather, it is necessary to examine students' behavioral patterns and psychological mechanisms to understand why they learn, how they learn, and what actions they take afterward.

This study is conducted against this backdrop. Drawing on Bandura's Social Cognitive Theory (SCT), it incorporates individual self-efficacy, coping strategies, and environmental factors into a unified framework to construct a driving model for BIM education effectiveness in construction management. Unlike prior research focusing primarily on teaching resources or curriculum design, this study emphasizes the triadic interaction of individual–environment–behavior. Specifically, environmental variables such as theoretical courses, project-based assignments, skill competitions, and institutional incentives are integrated with psychological variables such as career decision-making self-efficacy and coping efficacy, jointly influencing educational outcomes. A further innovation lies in the inclusion of both current students and industry professionals as comparative groups, enabling an analysis of differences in the underlying mechanisms. This heterogeneity perspective helps to uncover how learners at different career stages vary in motivational sources, efficacy development, and professional cognition, thereby providing evidence for differentiated educational interventions and talent development strategies.

Accordingly, this chapter seeks to address three key questions: first, how do university-level theoretical courses, practical activities, and institutional incentives influence students' career self-efficacy and coping mechanisms? Second, how do these psychological mechanisms, in turn, affect BIM education outcomes? Third, are there significant differences in these pathways between different groups? This study not only enriches the theoretical application of Social Cognitive Theory in engineering education but also provides empirical evidence for optimizing BIM curriculum design, enhancing practice-based teaching, and strengthening university–industry collaboration. More importantly, through a systematic analysis of the entire chain of "classroom–competition–career," it responds to three long-standing gaps in BIM education research: the lack of systematic modeling of learning motivation, insufficient examination of the educational value of competition contexts, and the neglect of comparative perspectives across different learner groups. By constructing an empirical model and conducting multi-group testing, this study aims to reveal the underlying logic of educational effectiveness and to provide actionable pathways for the continuous improvement of BIM education in Chinese universities and the effective alignment of talent cultivation with industry needs.

3.2 Literature Review

With the deepening application of Building Information Modeling (BIM) in the global architecture and construction management sectors, its integration into education has gradually become an important agenda in engineering education reform. A growing body of international research highlights that BIM is not merely a new digital tool but also a core methodological approach capable of reshaping engineering education models. Embedding BIM into university curricula helps cultivate students' comprehensive competencies, enabling them to acquire not only technical skills but also cross-disciplinary collaboration, information integration, and project management capabilities (Sacks & Barak, 2010). For example, Kolaric et al. (2017), in their evaluation of BIM education models in Europe, found that practice-oriented and project-driven teaching strategies significantly improved students' proficiency in BIM tools and their ability to apply them in real-world projects, underscoring the necessity of educational transformation.

In North America, BIM education started earlier and is closely integrated with industry practices. The National Institute of Building Sciences (NIBS), through its NBIMS-US, emphasized the importance of BIM education and promoted the systematic development of relevant curricula. Many universities adopt case-based and team collaboration teaching approaches to strengthen students' cross-disciplinary communication and decision-making skills (Lavi & Marti, 2023). In Australia and New Zealand, research pays more attention to aligning BIM education with industry standards, emphasizing that curriculum design must be consistent with employer needs to ensure graduates' smooth transition into the workforce (Casasayas et al., 2021; Guo et al., 2023).

In China, the development of BIM education has been strongly driven by national policies. The 2011–2015 Construction Industry Informatization Development Outline first set forth BIM promotion goals, followed by the Opinions on

Accelerating the Development of New-Type Construction Industrialization (2020) and the 14th Five-Year Plan, all of which explicitly positioned BIM as a core task. Against this backdrop, higher education institutions actively responded by incorporating BIM into the core curricula of engineering and architecture programs (Wang et al., 2017). Nevertheless, research indicates that, despite growing investment, the deep integration of theoretical courses with practice-oriented activities remains a major challenge. In many universities, BIM teaching remains limited to tool-level training, with insufficient project-based practice and interdisciplinary collaboration opportunities (Jin et al., 2020; Zhang et al., 2016).

Regarding the factors influencing educational effectiveness, both domestic and international scholars generally agree that teaching resources, practice environments, and university–industry collaboration constitute the three essential pillars for improving BIM education quality. International studies emphasize the importance of faculty training and interdisciplinary teams (Abdirad & Dossick, 2016), while Chinese research pays particular attention to infrastructure development and curriculum enhancement (Lyu et al., 2020). Some scholars also argue that only through close collaboration with industry experts can curricula remain dynamically updated and effectively aligned with industry demands (Kolaric et al., 2017; Zhang et al., 2016). These findings suggest that the effectiveness of BIM education depends not only on classroom instruction but also on systemic incentives, industry involvement, and the interplay of students' psychological motivations.

Nonetheless, the existing literature still exhibits several limitations. First, most studies focus on curriculum systems and teaching resources, with relatively little attention to students' learning behaviors and psychological mechanisms. Second, the educational value of non-formal learning pathways, such as BIM skill competitions, remains underexplored, with limited empirical research on their mechanisms and long-term effects. Third, current evaluation systems primarily rely on self-report surveys and subjective perceptions, lacking objective behavioral data to comprehensively capture educational effectiveness. Finally, many studies treat student populations as homogeneous, overlooking differences across learning stages or identities (e.g., current students versus industry professionals), which reduces the precision of educational intervention strategies.

To address these gaps, this study systematically reviews the domestic and international literature and attempts to construct a BIM education effectiveness driving model tailored to the Chinese higher education context. By integrating the analytical framework of Social Cognitive Theory, we link educational supply factors such as theoretical courses, practical activities, and institutional incentives with psychological factors such as career decision-making self-efficacy and coping efficacy. Furthermore, we compare two distinct groups—current students and industry professionals—to capture heterogeneity across populations. Through this design, we aim to identify the core drivers and mechanisms of BIM education effectiveness, thereby providing empirically grounded strategies and insights for optimizing university curricula, improving teaching methods, strengthening university–industry collaboration, and enhancing students' career adaptability.

3.3 Methods and Data

3.3.1 Theoretical Framework of the Study

3.3.1.1 Construction of the Basic Model

A substantial body of literature demonstrates that Social Cognitive Theory (SCT) has been widely applied in the field of higher education. Numerous scholars have examined the joint influence of individual and environmental factors on behavior. This study argues that the effectiveness of BIM education for both university students and industry professionals is similarly shaped by the interaction between individual attributes and environmental conditions.

In this research, BIM education effectiveness is defined as a behavioral variable, specifically referring to students' learning outcomes. These outcomes encompass the knowledge and understanding acquired through instructional activities (cognition), the development of attitudes and values (affect), the acquisition of practical skills (skills), and observable changes in behavior (Shi, 2012). For instance, students' attention to career information and development prospects related to BIM, as well as their proactive engagement in learning BIM technologies, are considered manifestations of BIM education effectiveness.

Accordingly, this study evaluates the effectiveness of BIM education by analyzing students' career choices, employment development, and engagement behaviors. The individual dimension refers to students' BIM-related career self-efficacy, including their cognitive perceptions of BIM technologies. The environmental dimension refers to the institutional context of BIM education within universities, the quality of which exerts a significant influence on educational effectiveness.

Based on these theoretical foundations, and in consideration of the specific characteristics of construction management programs and BIM education in higher education institutions, this study constructs a model of the factors influencing the effectiveness of BIM education (see Fig. 3.1).

3.3.1.2 Model Construction Based on Social Cognitive Theory

According to Social Cognitive Theory, behavior is jointly influenced by environmental and individual factors. In the context of this study, this means that the effectiveness of BIM education is shaped by both the institutional environment of BIM education and students' career self-efficacy in relation to BIM.

Therefore, grounded in this theoretical perspective, the study proposes a model of factors influencing the effectiveness of BIM education in construction management programs (see Fig. 3.2) and presents the overall research framework (see Fig. 3.3).

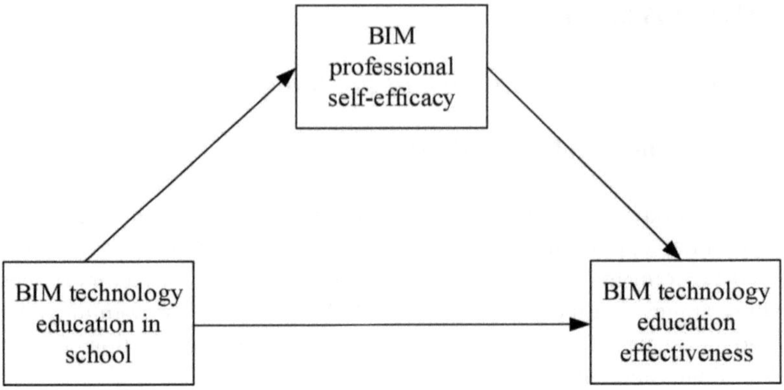

Fig. 3.1 Basic model of factors influencing the effectiveness of BIM education in construction management programs

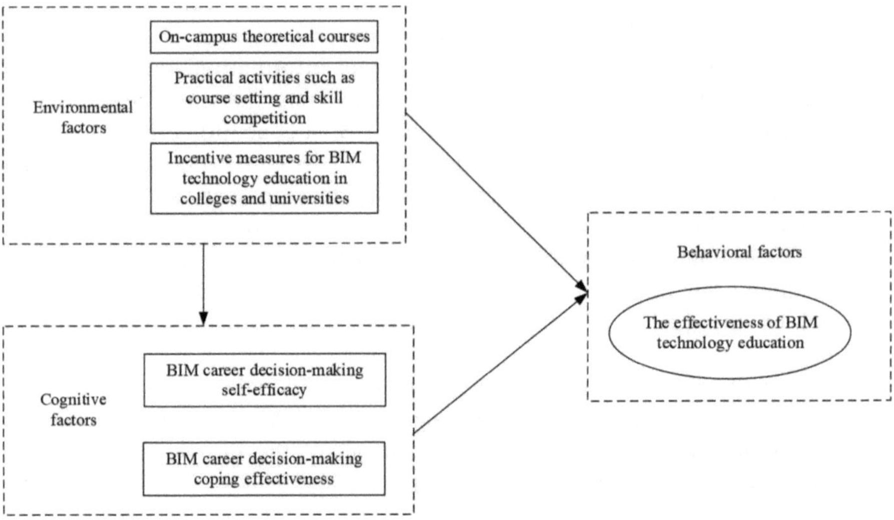

Fig. 3.2 Model of factors influencing the effectiveness of BIM education in construction management programs

3.3.1.3 Variable Design and Measurement

This study involves six latent variables, including theoretical courses in universities, practical activities (such as course projects and skills competitions), incentive measures for BIM education in higher education institutions, career decision-making self-efficacy in BIM, career decision-making coping efficacy in BIM, and effectiveness of BIM education. In total, 26 observed variables are included. Based on Social Cognitive Theory, this study adopts a five-point Likert scale for measurement, ranging from 1 (strongly disagree) to 5 (strongly agree).

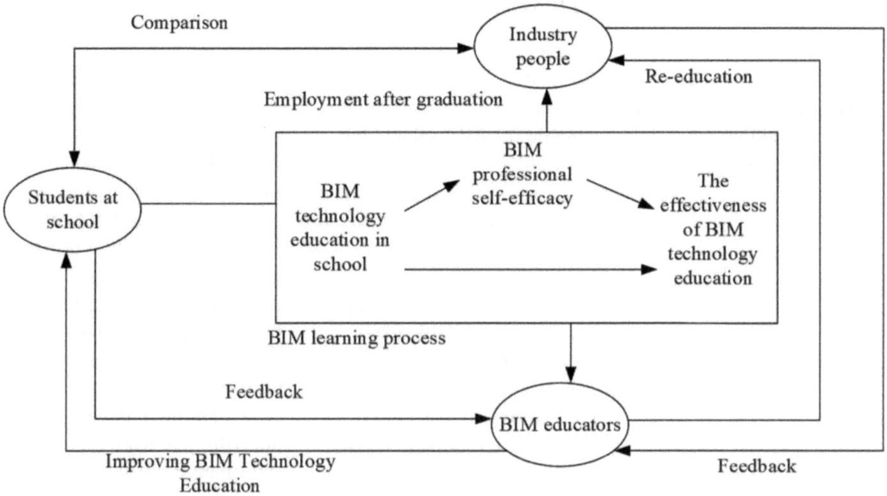

Fig. 3.3 Overall research framework of the study

1. Individual Factors

Bandura argued that in Social Cognitive Theory, individual behavior is determined by five capabilities, with self-efficacy at the core. He initially defined self-efficacy as an individual's confidence in their ability to complete a specific task. He later expanded this definition and explored how self-efficacy influences one's choices, effort, and persistence when facing difficulties. Self-efficacy reflects an individual's evaluation of their ability to successfully perform a specific task (Li et al., 2023). Higgins (1995) further divided self-efficacy into three dimensions: breadth, strength, and generality. Breadth refers to the perceived difficulty of a task, strength denotes the individual's confidence in their capability, and generality reflects their perception of ability in a specific context.

In this study, BIM career self-efficacy refers to students' evaluation of and confidence in their ability to successfully complete tasks related to BIM career exploration and decision-making. Lent et al. (2016) divided career decision-making self-efficacy into two dimensions: career decision-making self-efficacy and career decision-making coping efficacy. Accordingly, this study also subdivides BIM career self-efficacy into these two sub-variables.

Drawing on Betz and Luzzo (1996) Career Decision-Making Self-Efficacy Scale and Lent et al. (2016) short form of the Career Exploration and Career Decision-Making Self-Efficacy Scale, this study identifies measurement items for BIM career decision-making self-efficacy and BIM career decision-making coping efficacy. Considering the distinct characteristics of students and industry professionals, the measurement indicators are phrased differently for each group, as detailed in Tables 3.1 and 3.2.

Table 3.1 Measurement items for BIM career decision-making self-efficacy

Variable	Item code	Question	
		students	Industry professionals
BIM career decision-making self-efficacy (CDSE)	SE1	I am clear that my personality is suitable for BIM-related job positions	After graduation, I am clear that my personality is suitable for BIM-related job positions
	SE2	I am confident that BIM-related job positions can best utilize my skills	Since entering the workforce, I am confident that BIM-related job positions can best utilize my skills
	SE3	I believe that BIM-related jobs are the most suitable ideal careers for me	After entering the workforce, I believe that BIM-related jobs are the most suitable ideal careers for me
	SE4	I am willing to learn more information related to BIM technology	After entering the workforce, I am willing to learn more information related to BIM technology
	SE5	I find that my skills, values, and interests all align with BIM-related jobs	After entering the workforce, I find that my skills, values, and interests all align with BIM-related jobs
	SE6	I am able to make wise choices regarding BIM-related career options	Based on my current work experience, I am able to make wise choices regarding BIM-related career options
	SE7	I am willing to learn more knowledge related to BIM technology	Based on my current work experience, I am willing to learn more knowledge related to BIM technology
	SE8	I am certain that I am very interested in BIM-related jobs	After entering the workforce, I am certain that I am very interested in BIM-related jobs

Table 3.2 Measurement items for BIM career decision-making coping efficacy

Variable	Item code	Question
BIM career decision-making coping efficacy (CDCE)	CE1	If my first choice of a BIM-related job position is not successful, I can accept it calmly
	CE2	If my first choice of a BIM-related job position is not successful, I will find another career option and manage negative emotions
	CE3	If I realize that I cannot handle BIM-related work due to subjective reasons (e.g., insufficient personal skills), I will make an alternative plan
	CE4	If I realize that I cannot engage in BIM-related work due to objective reasons (e.g., downturn in the construction industry), I will make an alternative plan

In summary, the individual factors in the model include two latent variables: BIM Career Decision-Making Self-Efficacy and BIM Career Decision-Making Coping Efficacy.

2. Environmental Factors

According to Social Cognitive Theory, the environment not only influences the individual but also has a significant impact on individual behavior. This study focuses on BIM technology education within construction management programs, and therefore considers in-school BIM education as the primary environmental variable.

A review of existing literature indicates that despite the rapid development of the construction industry, there remains a mismatch between graduates' employment pressures and enterprises' demand for skilled professionals (Yin & Liu, 2013). This suggests a gap between construction management education and industry needs, highlighting the necessity for universities to adjust their curricula in line with market requirements. In-school theoretical courses play a crucial role in students' career development and employability (Zhang et al., 2020). Therefore, theoretical coursework is treated as a secondary latent variable in this study, given its potential importance in shaping students' employment decisions and career trajectories. Drawing on relevant domestic and international studies (Lucas, 2017; Wu & Issa, 2014), this study identifies four measurement items for in-school theoretical courses, as presented in Table 3.3.

Furthermore, studies have shown that skill competitions exert a significant positive impact on students' learning, employability, and career development (Wu, 2018). Encouraging students to participate in BIM-related competitions can effectively enhance their understanding of BIM (Yang & Wu, 2020). Numerous studies also confirm that incorporating BIM practice courses, such as course design and skill competitions, helps students strengthen their comprehension of BIM and better prepare for future employment and career development (LYU et al., 2020). Therefore, including skill competitions, course design, and other practice-oriented activities as secondary observed latent variables is justified, given their potential importance in fostering students' professional competencies.

Table 3.3 Measurement indicators for in-school theoretical courses

Variable name	Item code	Question
In-school theoretical courses (CE)	C1	Through my university theoretical courses, I have acquired the professional skills required for BIM-related work (e.g., software modeling, 5D construction simulation, etc.)
	C2	The BIM education system at my university is well established
	C3	The instructors of BIM-related courses possess strong professional competence and attach high importance to teaching
	C4	Through my university theoretical courses, I have developed a clear understanding of BIM-related job positions

Drawing on relevant studies (Zhang et al., 2020) and the actual educational context, this study identifies four measurement items for practice-oriented activities such as skill competitions and course design, as shown in Table 3.4.

In addition, motivational factors also exert a significant influence on individual behavior. According to Herzberg's Motivation Theory, motivational factors refer to elements that bring personal satisfaction and drive behavior (Guo, 2001). In the context of BIM education for engineering management majors, motivational factors are defined as a series of measures provided by universities to support BIM learning, such as BIM laboratories and related facilities, cross-disciplinary BIM education investment, BIM internships and cooperative projects, and inter-university BIM education collaboration.

According to Ao et al. (2022), strengthening BIM teaching facilities, providing adequate hardware and software support, deepening university–industry cooperation, improving foundational practice teaching platforms, and establishing specialized BIM training laboratories can effectively promote students' BIM learning behaviors. Similarly, Wu and Issa (2014) found that implementing practical and effective motivational measures can attract and encourage students to choose BIM-related jobs. Therefore, including university-level BIM educational incentives as a secondary observed latent variable is appropriate. Based on existing research and

Table 3.4 Measurement indicators for practice-oriented activities such as course design and skill competitions

Variable name	Item code	Question (Students)	
		Students	Industry professionals
Practice-oriented activities (PA)	PA1	I believe that completing BIM-related course design with high quality and participating in BIM skill competitions positively influences my professional understanding and confidence in pursuing a BIM career	I believe that completing BIM-related course design with high quality and participating in BIM skill competitions positively influences my professional understanding and confidence in pursuing a BIM career
	PA2	I believe that completing BIM-related course design with high quality and participating in BIM skill competitions positively impacts my employability and career development	I believe that completing BIM-related course design with high quality and participating in BIM skill competitions positively impacts my employability and career development
	PA3	I believe that construction enterprises fully recognize the role of BIM skill competitions and BIM-related course design in enhancing students' competence for BIM-related careers	I believe that construction enterprises fully recognize the role of BIM skill competitions and BIM-related course design in enhancing students' competence for BIM-related careers
	PA4	I believe that my university's BIM course design and skill competitions are highly relevant to the work I expect to engage in in the future	I believe that my university's BIM course design and skill competitions are highly relevant to the work I am currently engaged in

the actual context, this study identifies two measurement items for BIM educational incentives, as shown in Table 3.5.

In summary, the environmental variables in this model include three latent variables: in-school BIM education, practice-oriented activities such as course design and skill competitions, and university BIM educational incentives.

3. Behavioral Factors

In this study, behavioral factors are defined as the outcomes of BIM education, primarily reflected in changes in students' (or industry professionals') employment choices and career development behaviors after receiving BIM education. Defining BIM educational outcomes as a behavioral variable is particularly critical, as it not only serves as a key mechanism for students to enhance self-awareness and promote personal development, but also acts as an essential measure for universities to improve educational quality and achieve sustainable advancement (Shi, 2012). Based on relevant research and practical contexts, this study identifies four measurement items for BIM educational outcomes, as shown in Table 3.6.

3.3.2 *Questionnaire Design Framework*

This study designed two versions of the questionnaire, targeting current students and industry professionals respectively. Each version was divided into three sections: The first section collected basic demographic information. The student version included gender, university, grade level, intended employment sector, and expected profession's relevance to BIM technology. The industry professional version included gender, alma mater, years since graduation, industry sector of the current employer, relevance of the current position to BIM technology, and current job title.

The second section investigated the general status of BIM education in construction management programs, covering BIM learning content, deficiencies in university BIM education, and the career development incentives provided by universities.

The third section comprised the measurement scales for factors influencing BIM educational outcomes in construction management programs. Respondents were asked to rate the degree to which each statement matched their own situation using a five-point Likert scale (from 1 = strongly disagree to 5 = strongly agree).

Table 3.5 Measurement indicators for university BIM educational incentives

Variable Name	Item code	Question
University BIM Educational Incentives (IM)	IM1	I believe that my university's BIM educational incentives are comprehensive.
	IM2	I believe that my university's BIM educational incentives are highly effective.

Table 3.6 Measurement indicators for BIM educational outcomes

Variable name	Item code	Question	
		Students	Industry professionals
BIM educational outcomes (DB)	DB1	I continuously keep track of the latest news and developments in BIM-related work	Since starting work, I have continuously kept track of the latest news and developments in BIM-related work
	DB2	I actively participate in BIM skill competitions and related skill learning	Since starting work, I have actively participated in BIM training and industry competitions
	DB3	I pay attention to the career prospects and future trends of BIM-related jobs	Since starting work, I have paid attention to the career prospects and future trends of BIM-related jobs
	DB4	I plan to apply for BIM-related job positions and pursue a long-term career in BIM-related industries	Since starting work, I have followed multiple BIM-related projects and pursued in-depth development in BIM-related technologies

This study addressed factors including in-school BIM education, BIM career self-efficacy, and BIM educational outcomes. Six latent variables were designed: in-school theoretical courses, practical activities (e.g., coursework and skill competitions), university BIM education incentives, BIM career decision-making self-efficacy, BIM career decision-making coping efficacy, and BIM educational outcomes. The questionnaire items were differentiated in wording for the two respondent groups. To ensure reliability and validity, the scales referenced well-established measures from prior domestic and international empirical studies, while incorporating the practical context of BIM education in construction management.

After multiple rounds of testing and revision, including a pilot survey with over twenty faculty, students, and alumni via online communication, the research team refined the wording of each item for clarity and ease of understanding. Ultimately, a measurement scale of 26 items was finalized for BIM educational outcomes in construction management programs, as shown in Table 3.7.

3.3.3 3Data Collection

The questionnaire survey and data collection for this study began in April 2023. Prior to the formal survey, a pilot study was conducted to thoroughly revise the questionnaire items, ensuring their accuracy and applicability. As a result, two separate versions of the questionnaire were finalized—one targeting current students and the other targeting industry professionals—in order to meet the needs of different respondent groups.

The primary data collection focused on universities and colleges in the Sichuan–Chongqing region, covering both current students in construction management programs and graduates already employed in the industry. Institutions offering

Table 3.7 Measurement scale for factors influencing BIM educational outcomes in construction management programs

Factor category	Latent variable	Code	Measurement items	
			Students	Industry professionals
Environmental factors	In-school theoretical courses (CE)	C1	Through university theoretical courses, I have mastered the professional skills required for BIM-related work (e.g., software modeling, 5D construction simulation)	Through university theoretical courses, I have mastered the professional skills required for BIM-related work (e.g., software modeling, 5D construction simulation)
		C2	The BIM education system at my university is well established	The BIM education system at my university was well established
		C3	The instructors of BIM-related courses have strong professional competence and attach high importance to teaching	The instructors of BIM-related courses had strong professional competence and attached high importance to teaching
		C4	Through university theoretical courses, I have gained a clear understanding of BIM-related job positions	Through university theoretical courses, I gained a clear understanding of BIM-related job positions
	Coursework, skill competitions, and other practical activities (PA)	PA1	I believe that successfully completing BIM-related coursework and participating in BIM skill competitions positively impacts professional understanding and confidence in pursuing BIM careers	I believe that successfully completing BIM-related coursework and participating in BIM skill competitions positively impacts professional understanding and confidence in pursuing BIM careers
		PA2	I believe that successfully completing BIM-related coursework and participating in BIM skill competitions positively impacts employment and career development	I believe that successfully completing BIM-related coursework and participating in BIM skill competitions positively impacts employment and career development
		PA3	I believe that construction companies fully recognize the role of BIM skill competitions and BIM-related coursework in enhancing students' professional competence	I believe that construction companies fully recognize the role of BIM skill competitions and BIM-related coursework in enhancing students' professional competence
		PA4	The BIM coursework and skill competitions at my university are highly relevant to my future job	The BIM coursework and skill competitions at my university are highly relevant to my current job
	University BIM education incentives (IM)	IM1	I believe my university's BIM education incentives are comprehensive	I believe my university's BIM education incentives were comprehensive
		IM2	I believe my university's BIM education incentives are highly effective	I believe my university's BIM education incentives were highly effective

(continued)

Table 3.7 (continued)

Factor category	Latent variable	Code	Measurement items	
			Students	Industry professionals
Cognitive factors	BIM career decision-making self-efficacy (CDSE)	SE1	I am aware that my personality is suitable for BIM-related job positions	After starting work, I realized that my personality is suitable for BIM-related job positions
		SE2	I am confident that BIM-related job positions can best utilize my skills	Since working, I have confirmed that BIM-related job positions can best utilize my skills
		SE3	I believe BIM-related work is the ideal career path for me	Since working, I believe BIM-related work is the ideal career path for me
		SE4	I am willing to learn more about BIM-related information	Since working, I have been willing to learn more about BIM-related information
		SE5	I have found that my skills, values, and interests align well with BIM-related work	Since working, I have found that my skills, values, and interests align well with BIM-related work
		SE6	I can make informed choices regarding BIM-related career options	With my current work experience, I can make informed choices regarding BIM-related career options
		SE7	I am willing to learn more knowledge about BIM technology	With my current work experience, I am willing to learn more knowledge about BIM technology
		SE8	I am certain that I am very interested in BIM-related jobs	Since working, I have been certain that I am very interested in BIM-related jobs
	BIM career decision-making coping efficacy (CDCE)	CE1	If my first-choice BIM-related job position does not work out, I can accept it calmly	If my first-choice BIM-related job position does not work out, I can accept it calmly
		CE2	If my first-choice BIM-related job position does not work out, I will find another option and manage negative emotions	If my first-choice BIM-related job position does not work out, I will find another option and manage negative emotions
		CE3	If I realize I cannot handle BIM-related work due to subjective reasons (e.g., insufficient skills), I will make backup plans	If I realize I cannot handle BIM-related work due to subjective reasons (e.g., insufficient skills), I will make backup plans
		CE4	If I realize I cannot pursue BIM-related work due to objective reasons (e.g., a downturn in the construction industry), I will make backup plans	If I realize I cannot pursue BIM-related work due to objective reasons (e.g., a downturn in the construction industry), I will make backup plans

(continued)

Table 3.7 (continued)

Factor category	Latent variable	Code	Measurement items: Students	Measurement items: Industry professionals
Behavioral factors	BIM educational outcomes (DB)	DB1	I continuously keep track of the latest BIM-related news and developments	Since working, I have continuously kept track of the latest BIM-related news and developments
		DB2	I actively participate in BIM-related skill competitions and learning activities	Since working, I have actively participated in BIM training and industry competitions
		DB3	I pay attention to the career prospects and development trends of BIM-related jobs	Since working, I have paid attention to the career prospects and development trends of BIM-related jobs
		DB4	I plan to apply for BIM-related job positions and pursue a long-term career in the BIM-related industry	Since working, I have followed multiple BIM-related projects and pursued in-depth development in BIM-related technologies

construction management (including cost engineering and building information tracks) at both undergraduate and vocational levels were randomly selected as survey targets. Given that core courses in construction management programs are typically concentrated in the later years of study, third- and fourth-year undergraduates, as well as third-year vocational students, were chosen as the primary respondents.

During the sampling process, we established direct contacts with universities and excluded those with weak willingness to cooperate. The final list of sample institutions participating in data collection is presented in Table 3.8. In addition, faculty members from these institutions were contacted to distribute questionnaires to recent graduates and to extend the survey through their employers, thereby collecting data from industry professionals.

The formal questionnaire survey was conducted from April 13 to April 24, lasting a total of 12 days. Data collection progress was recorded daily at 10:00 a.m., as summarized in Table 3.9. In total, 721 student questionnaires and 202 industry professional questionnaires were collected.

To ensure the reliability of responses, both versions of the questionnaire included a control question (item 26) serving as a lie-detection item, requiring respondents to select "1." Only questionnaires with correct responses to this item were considered valid. After excluding invalid responses, 351 valid student questionnaires (validity rate: 48.8%) and 125 valid industry professional questionnaires (validity rate: 61.9%) were retained for analysis.

Table 3.8 List of sample institutions for the survey

Region	Institution	Type
Sichuan	Chengdu University of Technology	Research-oriented, applied
Sichuan	Sichuan College of Architectural Technology	Applied
Sichuan	SWUFE Tianfu College	Private
Sichuan	Chengdu University	Applied
Sichuan	Xihua University	Applied
Chongqing	Chongqing Technology and Business University	Research-oriented, applied
Chongqing	Chongqing Jiaotong University	Research-oriented, applied
Sichuan	Southwest Petroleum University	Research-oriented, applied
Sichuan	Southwest Jiaotong University Hope College	Private
Sichuan	Chengdu University of Technology and Industry	Applied

Table 3.9 Reliability analysis results

Variable	Items	Cronbach's alpha	KMO	Bartlett's P
In-school theoretical courses	4	0.900		
Course design, skills competitions, and other practical activities	4	0.910		
Incentive measures for BIM education in universities	2	0.952		
BIM career decision-making self-efficacy	8	0.949		
BIM career decision-making coping efficacy	4	0.904		
BIM education effectiveness	4	0.927		
Total	26	0.964	0.947	0.000

3.4 Research Results

3.4.1 Reliability and Validity Tests

This study employed SPSS 27.0 for data analysis. The Cronbach's Alpha coefficient was 0.964, indicating a high level of internal consistency across the questionnaire. Each subscale showed Cronbach's Alpha values above 0.8, meeting reliability requirements. The KMO value was 0.947, significantly exceeding the 0.7 threshold, and Bartlett's test of sphericity produced a p-value of 0, confirming that the data were suitable for factor analysis.

3.4.2 Model Fit

SmartPLS was applied for structural equation modeling (SEM). The results showed that the coefficient of determination (R^2) values ranged between 0.3 and 0.7, and the SRMR value was 0.072, both within acceptable standards. The Q^2 values were greater than 0, indicating good predictive relevance of the model. The GoF index was 0.57, higher than the 0.36 threshold, suggesting that the overall model fit was satisfactory (Table 3.10).

3.4.3 *Heterogeneity in the Empirical Testing of the Driving Model*

Using permutation multigroup analysis and bootstrap multigroup analysis in SmartPLS, this study compared students and industry professionals. Results indicated no significant differences in hypothesis testing outcomes between the two groups. Hypotheses H1 to H6 were supported, while H7 and H8 were not.

In the path analysis, neither group showed significant direct effects of theoretical courses, incentive measures, or practical activities on BIM education effectiveness. However, BIM career decision-making coping efficacy exerted a very significant positive influence on education effectiveness for both groups, with stronger effects among industry professionals than students. The influence of theoretical courses and incentive measures on coping efficacy was also greater among industry professionals, whereas practical activities had a stronger impact on coping efficacy among students.

Mediation effect analysis further revealed that the indirect effect of practical activities on BIM education effectiveness via coping efficacy was more pronounced among students than industry professionals. Similarly, the indirect effects of theoretical courses and incentive measures on education effectiveness via self-efficacy were very significant in both groups, with comparable effect sizes (Tables 3.11, 3.12, and 3.13).

3.5 Discussion

Through questionnaire surveys and structural equation modeling analyses of both students and industry professionals, this study examined the general state of BIM education in construction management programs, identified the driving factors of

Table 3.10 Model fit indicators

Latent variable	R^2 (>0.1, higher is better)	SRMR (<0.1 preferred)	Q2(>0)	AVE	GoF
CDCE	0.326	0.072	0.230	0.704	0.57
CDSE	0.362		0.243	0.701	
CEIM	–		–	0.717	
DB	0.644		0.466	0.760	
PA	–			0.780	
Average value	0.444		–	0.732	

Table 3.11 Hypothesis testing results for student and industry professional groups

Hypothesis	Student group	Industry professional group
H1: Theoretical courses and incentive measures have a significant positive impact on BIM career decision-making self-efficacy	β = 0.403, p = 0.000, supported	β = 0.423, p = 0.000, supported
H2: Theoretical courses and incentive measures have a significant positive impact on BIM career decision-making coping efficacy	β = 0.124, p = 0.043, supported	β = 0.245, p = 0.032, supported
H3: Practical activities (course design, skills competitions) have a significant positive impact on BIM career decision-making coping efficacy	β = 0.452, p = 0.000, supported	β = 0.245, p = 0.032, supported
H4: Practical activities have a significant positive impact on BIM career decision-making self-efficacy	β = 0.324, p = 0.000, supported	β = 0.247, p = 0.008, supported
H5: BIM career decision-making self-efficacy has a significant positive effect on BIM education effectiveness	β = 0.707, p = 0.000, supported	β = 0.651, p = 0.000, supported
H6: BIM career decision-making coping efficacy has a significant positive effect on BIM education effectiveness	β = 0.182, p = 0.001, supported	β = 0.391, p = 0.000, supported
H7: Practical activities have a significant positive effect on BIM education effectiveness	β = −0.094, p = 0.069, not supported	β = −0.053, p = 0.539, not supported
H8: Theoretical courses and incentive measures have a significant positive effect on BIM education effectiveness	β = 0.031, p = 0.554, not supported	β = −0.174, p = 0.088, not supported

educational effectiveness, and explored their heterogeneous pathways. The findings highlight the heterogeneity between the two groups in terms of cognition, motivation, and educational needs, providing new empirical evidence for understanding the mechanisms of BIM education.

First, regarding perceptions of the general state of BIM education, there were no significant differences between students and industry professionals. This suggests that both groups, whether situated in academic or industry contexts, generally acknowledge the importance of BIM education. However, when examining deficiencies and course design in detail, clear divergences emerged. Industry professionals emphasized the lack of project experience, insufficient practical courses, and teachers' limited real-world engineering backgrounds—issues directly tied to their practical need for "applicability and usability." In contrast, students stressed the lack of multidisciplinary collaboration and insufficient hands-on training, reflecting their immediate experiences within the academic environment, where they primarily encounter classroom teaching and simulated tasks. Put differently, industry professionals assess deficiencies from the perspective of job readiness, whereas students express concerns from their learning process experience.

Second, differences also arose in the understanding and expectations of course content. Industry professionals expected BIM education to cover a wide range of aspects—including modeling, schedule control, site management, and cost

Table 3.12 Path analysis indicators for student and industry professional groups

Path	Path coefficient (A)	Path coefficient (B)	T(A)	T(B)	P(A)	P(B)	Difference (A-B)	2-tailed (AVB)p
CDCE -> DB	0.182	0.391	3.404	4.183	0.001	0.000	−0.210	0.048
CDSE -> DB	0.707	0.651	13.331	7.958	0.000	0.000	0.056	0.576
CEIM -> CDCE	0.124	0.391	2.024	3.232	0.043	0.001	−0.267	0.061
CEIM -> CDSE	0.403	0.423	7.544	4.404	0.000	0.000	−0.020	0.840
CEIM -> DB	0.031	−0.174	0.592	1.708	0.554	0.088	0.205	0.071
PA -> CDCE	0.452	0.245	7.465	2.141	0.000	0.032	0.207	0.116
PA -> CDSE	0.324	0.247	5.727	2.638	0.000	0.008	0.078	0.473
PA -> DB	−0.094	−0.053	1.822	0.615	0.069	0.539	−0.040	0.671

Table 3.13 Mediation effects for student and industry professional groups

Path	Path coefficient (A)	Path coefficient (B)	T(A)	T(B)	P(A)	P(B)	Difference (A-B)	2-tailed (AVB)p
PA -> CDCE -> DB	0.229	0.161	5.067	2.474	0.000	0.013	0.069	0.381
CEIM -> CDCE -> DB	0.023	0.153	1.711	2.664	0.087	0.008	−0.131	0.015
CEIM -> CDSE -> DB	0.285	0.275	6.823	3.765	0.000	0.000	0.009	0.903
PA -> CDSE -> DB	0.082	0.096	3.149	1.741	0.002	0.082	−0.014	0.886

estimation—so as to align more effectively with the integrated demands of real-world projects. Students, however, focused more heavily on modeling skills. This divergence indicates that current curricula still lean toward a "tool-centered, management-light" approach, underestimating BIM's systemic value as a platform for project integration and collaboration. Such imbalance not only limits students' holistic understanding of BIM but also contributes to a mismatch between their skills and job requirements once they enter the workforce.

Third, the pathways driving educational effectiveness differed between the two groups. For students, practical activities such as course projects and skills competitions exerted stronger positive influences on career self-efficacy and overall educational outcomes, underscoring the crucial role of practice-based teaching in shaping student capabilities. This finding echoes the "situated learning" perspective in educational psychology, which posits that learners more effectively construct knowledge and develop self-efficacy in authentic or near-authentic task environments. Conversely, for industry professionals, educational outcomes were more strongly influenced by theoretical courses and institutional incentives, likely because their existing work experience reduced reliance on skills training, while theory and institutional support played a greater role in shaping career confidence and development strategies.

Further analysis revealed distinct heterogeneity in the influence of career self-efficacy and coping efficacy. Students were more strongly affected by practical activities, suggesting that universities should further strengthen project-based courses, hands-on training, and competitions to help students form clearer career orientations during their academic years. Industry professionals, however, placed greater emphasis on theoretical courses and institutional incentives, signaling the need for universities to strengthen ties with industry, keep curricula aligned with evolving professional standards, and enhance career confidence and adaptability through initiatives such as industry mentorships, internships, and professional certification systems.

Moreover, this study highlighted a misalignment between the supply of BIM education and industry demand. Students placed more importance on the learning experience itself, while industry professionals prioritized job relevance and practical competence. This gap suggests that curricula should not only focus on the "classroom-to-competition" pathway but also systematically incorporate industry needs into learning objectives. Strategies such as bringing in industry mentors, simulating real-world projects, and fostering interdisciplinary teamwork could help bridge this divide and maximize educational effectiveness.

In summary, this discussion offers several key implications: (1) universities should strike a balance between theory and practice in BIM teaching, ensuring both systematic foundations and rich practical exposure; (2) stronger industry-academia collaboration is needed, with greater involvement of practitioners to align curricula with industry standards and job requirements; (3) multidimensional evaluation systems should be established, integrating learning motivation, competition performance, and employability into comprehensive feedback mechanisms; and (4) differentiated educational needs between students and industry professionals should be addressed through targeted teaching strategies and group-based approaches to ensure precise allocation of educational resources.

By comparing students and industry professionals, this study not only identifies the key factors influencing BIM education effectiveness and their pathways of action but also provides empirical support for understanding the dynamic relationship between education and industry. These findings carry significant practical implications for improving BIM education quality in Chinese construction management programs, optimizing talent development models, and meeting the workforce demands of the industry's digital transformation.

3.6 Conclusion

This study focused on BIM education in construction management programs, developing a performance-driven model based on Social Cognitive Theory (SCT) and conducting empirical analyses across two groups: students and industry professionals. The research not only verified the multidimensional structure of educational effectiveness through the design of measurement scales and structural equation modeling but also revealed the practical challenges of integrating theory with practice in BIM education.

The main contributions of this study can be summarized in three aspects. First, at the theoretical level, it incorporated psychological variables such as self-efficacy and coping efficacy into the framework for analyzing BIM educational outcomes, thereby enriching the explanatory power of engineering education research. Second, at the methodological level, the study developed and validated a measurement scale consisting of 26 items, offering subsequent researchers a tool that can be directly applied or further extended. Third, at the applied level, the study not only demonstrated the causal logic by which educational supply factors (courses, practice, incentives) influence outcomes, but also included the perspectives of industry professionals, thereby creating a bridge for dialogue between academia and industry.

The findings of this study provide practical implications for the reform of BIM education in higher education institutions. Universities should strive to balance theory and practice within curricula, increase resource investment, and strengthen industry-academia collaboration. In terms of educational assessment, they should move beyond traditional grade-oriented approaches and establish comprehensive evaluation frameworks that encompass learning motivation, career efficacy, and industry adaptability. Regarding talent cultivation, greater attention should be given to the entire trajectory from learning to career development, ensuring that education simultaneously serves both industry transformation and student growth.

Overall, this study not only systematically presents the effectiveness and challenges of BIM education but also makes exploratory contributions at the theoretical, methodological, and practical levels. Although there remains room for improvement, the findings provide a solid reference for improving BIM education quality in Chinese universities, optimizing pathways for talent development, and promoting deeper integration between education and industry.

References

Abdirad, H., & Dossick, C. S. (2016). BIM curriculum design in architecture, engineering, and construction education: a systematic review. *Journal of Information Technology in Construction, 21*(17), 250–271.

Ao, Y., Peng, P., Li, J., Li, M., Bahmani, H., & Wang, T. (2022). What determines BIM competition results of undergraduate students in the architecture, Engineering and Construction Industry? *Behavioral Science, 12*(10), 360. https://doi.org/10.3390/bs12100360

Betz, N. E., & Luzzo, D. A. (1996). Career assessment and the career decision-making self-efficacy scale. *Journal of Career Assessment, 4*(4), 413–428.

Casasayas, O., Hosseini, M. R., Edwards, D., Shuchi, S., & Chowdhury, M. (2021). Integrating BIM in higher education programs: Barriers and remedial solutions in Australia. *Journal of Architectural Engineering, 27*(1), 05020010.

Guo, H. (2001). Review of incentive theory. *Enterprise Economy*, (06), 32–34. https://kns.cnki.net/kcms2/article/abstract?v=HIS3QibhfQrekHD_sbxUBTgcB6Ei79kabDZPfnQHly-D3x89gY4DkMwZA8uLhkmWJhWXlCzs8N9uVmbL3HEeeDTBSFh198GrEYbs4Jws-kOnUU_GDIHOxGraqoWOh1CWK8CFU4xjUCuPc_S9JM1ycuhME1EBFV0nShF7_wBcuTU5fLRDJle7C8jA==&uniplatform=NZKPT&language=CHS

Guo, B. H., Gonzalez, V. A., Puolitaival, T., Enegbuma, W., & Zou, Y. (2023). Bridging the gap between building information modelling education and practice: a competency-based education perspective. *International Journal of Construction Management, 23*(15), 2558–2569.

Higgins, C. C. A. (1995). Computer self-efficacy: Development of a measure and initial test. *MIS Quarterly, 19*(2), 189–211.

Jin, J., Hwang, K. E., & Kim, I. (2020). A study on the constructivism learning method for BIM/IPD collaboration education. *Applied Sciences-Basel, 10*(15), article 5169. https://doi.org/10.3390/app10155169

Kolaric, S., Vukomanovic, M., Stober, D., & Dolacek-Alduk, Z. (2017). Assessing educational approaches to building information modelling (bim) at construction management master studies in Croatia. *Tehnicki Vjesnik-Technical Gazette, 24*(4), 1255–1262. https://doi.org/10.17559/tv-20160922083031

Lavi, R., & Marti, D. (2023). A proposed case-based learning framework for fostering undergraduate engineering students' creative and critical thinking. *Journal of Science Education and Technology, 32*(6), 898–911.

Lent, R. W., Ezeofor, I., Morrison, M. A., Penn, L. T., & Ireland, G. W. (2016). Applying the social cognitive model of career self-management to career exploration and decision-making [Article]. *Journal of Vocational Behavior, 93*, 47–57. https://doi.org/10.1016/j.jvb.2015.12.007

Li, S. Y., Hong, Y. C., & Craig, S. D. (2023). A systematic literature review of social learning theory in online learning environments. *Educational Psychology Review, 35*(4), 108. https://doi.org/10.1007/s10648-023-09827-0

Lucas, J. D. (2017). Identifying learning objectives by seeking a balance between student and industry expectations for technology exposure in construction education [Article]. *Journal of Professional Issues in Engineering Education and Practice, 143*(3), 05016013. https://doi.org/10.1061/(asce)ei.1943-5541.0000318

Lyu, W., Zhang, S., Jiang, R., & Shane, G. (2020). Current situation of competitiveness cultivation of engineering management graduates based on empirical analysis of Suzhou. *Project Management Technology, 18*(08), 49–54. https://kns.cnki.net/kcms2/article/abstract?v=HIS3QibhfQqQvJbndN8fJTuQYvH_9QAQBwwTbtVs-NijTVkZhcnJfciCeZtby4h6ILFu6o36S2QXFUuOPWQEKgXUFlUp00X0UikCW5zDBmSu-cz2wVnVpnPCxgMntfhty2NI5_AfAZqj28FgDo33-xra4bYFXF4sxC-UIgcNL9_2PetXgHNTMTTqORIqMxfu&uniplatform=NZKPT&language=CHS

Sacks, R., & Barak, R. (2010). Teaching building information modeling as an integral part of freshman year civil engineering education. *Journal of Professional Issues in Engineering Education and Practice, 136*(1), 30–38. https://doi.org/10.1061/(ASCE)EI.1943-5541.000000

Shi, J. (2012). *Research on learning effectiveness evaluation of undergraduates in research universities* (Master). https://kns.cnki.net/kcms2/article/abstract?v=HIS3QibhfQrqn4nOAcCWX2W1BxwqpMD4QdgkaMQ3NN1GX2K5cpwn4IE1bBtqkwMo1-DvfPmRKfEcJ0xyya6onnkGAaMd7su977KSvMZStzQcZmdTCFPE8XBoUsSbU7L0Q-mQJDeRufX8Zj0bOzst72fQa7XtgPIZXNUy2hhQCxZphSUS1FrbEdNbqzSMY1z9&uniplatform=NZKPT&language=CHS

Wang, J., Zhang, D., & Zhou, J. (2017). Exploration on the application of BIM technology in the teaching of professional courses in architectural colleges and universities – Taking Shenyang Jianzhu university as an example. *Journal of Architectural Education in Institutions of Higher Learning, 26*(01), 161–164. https://kns.cnki.net/kcms2/article/abstract?v=Eh6WGTKUEQTq-btgG8GgHBLeVzAtCAkQv0kHVGe40xqf3NoSiw17Xp8wpJcEs8Vt361flvUXxw0ap8lj_GT7MqPnpHODpmEfxwc0YPpgXMwtV-eI9B7QGYtsMs7ojAVpGlxFP-KJHgReysSA0HGQp3CwMSze0bMrcSZ16swudrUzh9jIo94CO_odngrcQMPtAYBFOi-iHw34=&uniplatform=NZKPT&language=CHS

Wu, Y. (2018). Analysis of the influence of skill competition on the career development of higher vocational graduates. *Journal of Jiyuan Vocational and Technical College, 17*(03), 45–49. https://kns.cnki.net/kcms2/article/abstract?v=HIS3QibhfQozl408AgFJHPf2G6WuNXRdkSCvL0Gow2c_iGp9wuLo7rGJCjTKwd5UV217JwtfmcLiIH9oE1SkJggQIlNe4yXv-DNnXFXqio2yASUGXDHhNjIQH14pvXBHpFEOuWFPRGV1WuT7zaqw554OPH40viWK9ZCoeDt301sKYhMK_LtoquqJRr5bxKxW&uniplatform=NZKPT&language=CHS

Wu, W., & Issa, R. R. A. (2014). BIM education and recruiting: Survey-based comparative analysis of issues, perceptions, and collaboration opportunities [Article]. *Journal of Professional Issues in Engineering Education and Practice, 140*(2). https://doi.org/10.1061/(asce)ei.1943-5541.0000186

Yang, J., & Wu, Y. (2020). Discussion on the professional development after the integration of engineering management specialty and BIM. *Science & Technology Ecnony Market, 06*, 133–134. https://kns.cnki.net/kcms2/article/abstract?v=HIS3QibhfQr-HbW0VCu3NdXzH9HMDWE57zSR4dd-YPSXGydXT7YAq-st89tjVzL-cE4GzfbbBvqCANiF55nN7bQg0o7TEocQoKy_A2GyK9rlQbhSAE9r-2G0_20vQcBiLWT8anxIx2lY2eeJ9lmIMEJNADsHIJoZMPcXurM0d93eje-ipNCk528VDY9ltMHE&uniplatform=NZKPT&language=CHS

Yin, Y., & Liu, G. (2013). The exploration and research on the assessment of professional competence of application-oriented undergraduate – A Case study of construction management specialties. *Modern Educational Technology, 23*(11), 119–123+189. https://kns.cnki.net/kcms2/article/abstract?v=HIS3QibhfQr32YiSPQ1XTJCkzb1rbNoeyhoj93qnDZGcqDWYsJUaDtOhIx6EsGONBZIfT6Hc8EbJYeVKqDaSPbtms0fHci_g4hK1F6GdrGuRqO6PEcbu0eCAnMpargfKHaLtWFizI7aVFNXx2jhKdxZB5VC1QESqizPmtgKctvd4-66wUUKDyA==&uniplatform=NZKPT&language=CHS

Zhang, J. X., Xie, H. Y., & Li, H. (2016). Exploring the cognitive structure and quality elements: Building information modeling education in civil engineering and management. *International Journal of Engineering Education, 32*(4), 1679–1690. <Go to ISI>://WOS:000389147200014.

Zhang, H., Tang, G., & Ding, H. (2020). Path of advancing Core competence of engineering management major – Based on the perspective of professional development needs. *Journal of Beibu Gulf University, 35*(08), 59–65. https://doi.org/10.19703/j.bbgu.2096-7276.2020.08.0059

Chapter 4
Factors Impacting BIM Educational Effectiveness: Evidence from Construction Management Students

Abstract Given the widespread application of BIM technology across the construction industry, there is a commensurate need to cultivate professionals able to fulfil industry expectations. In particular, assessing the current effectiveness of BIM education within construction management programs at universities is a necessary step in delivering graduates sufficiently adept in BIM practice. In this study, we constructed a model of the driving factors of BIM education effectiveness in the field of construction management based on Bandura's Social Cognitive Theory (SCT). Data was drawn from 351 valid, returned questionnaires from an initial 721 distributed to candidate respondents across the Sichuan-Chongqing region, China. The results revealed that the direct impact of both the theoretical BIM educational environment and practical BIM education quality on BIM education effectiveness was not significant. They nevertheless had a significant positive indirect impact through the generation of BIM professional self-efficacy. The influence of BIM professional decision-making self-efficacy proved to be greater than that of BIM professional decision-making coping self-efficacy. Based on these findings, this study provides targeted recommendations for further enhancing the effectiveness of BIM education in the field of engineering management, both within China and elsewhere.

Keywords BIM education · Construction management · Social cognitive theory · Self-efficiency · Career development

4.1 Introduction

Building Information Modeling (BIM) technology has become an indispensable tool used globally in the field of construction, the use of which is often mandated at national and local levels. This is the case in China. Similarly, the education sector

Y. Ao et al., *Building Information Modeling (BIM) Adoption*, Resilient and Friendly Living Environments, https://doi.org/10.1007/978-981-95-3993-2_4

has also fully recognized the importance of BIM technology (Li et al., 2020). However, graduates continue to demonstrate a persistent insufficient skill set in BIM, which in turn has hindered its application within industry (Sacks & Barak, 2010). Consequently, the cultivation and lifting of BIM educational outcomes has emerged as a priority within the construction industry (Castro-Lacouture et al., 2014). There is continuing interest in the factors impacting BIM educational effectiveness. Based on social cognitive theory, our research constructs a model of influencing factors of BIM education effectiveness for engineering management majors, aiming to explore the influence of BIM education on the formation of BIM career self-efficacy and the final career choice and development. In China, the rapid development of industrial technology has placed significant demands on senior professionals (Gu, 2004; Li, 2018). Keeping abreast of BIM benchmark practices in the construction industry and recruiting and retaining personnel able to manage projects utilizing BIM is one of the most significant challenges faced (Wu et al., 2021). Significant research has already been undertaken to identify feasible ways to incorporate BIM technology into the engineering undergraduate education system in universities in China. These include analyzing specific reform plans for training programs in engineering management majors, and identifying obstacles to the application of BIM technology in the teaching of civil construction majors (Ao et al., 2022; Ma & Tao, 2023; Peng et al., 2022). Foreign scholars have conducted more in-depth discussions on BIM education and BIM practice, for example, the case study of Madrid Polytechnic University shows that the reasons students postpone the learning of BIM-related tools have to do with the lack of rules and regulations, the standard operating procedures, and the lack of profitability during the period of use (Lopez-Zaldivar et al., 2017). Lucas (2017) points out that in order to ensure that students graduating from construction programs are prepared to serve the industry, it is necessary for universities to include construction technology in their curricula and conducted a study on teaching programs for BIM courses designed to meet students' BIM expectations while satisfying the industry's needs. Patrick X. W. Zou et al. (2019) conducted a study on BIM students' perceptions of BIM practice, providing insights for BIM educators and AEC employers in optimizing BIM educational resources between industry practice and academic research, understanding grouping differences in their perceptions and motivations, and similarities and differences between AEC students and industry practitioners. Suchi et al. (2020) produced a report for the BIM Academic Forum, concluding that the main barriers to BIM progress were lack of in-house expertise, lack of training, and lack of client demand. Similarly, Martek et al. (2021) concluded that while current university BIM education does expose students to the raw fundamentals of BIM software usage, this was insufficient in translating to utilization of BIM as a problem-solving platform in real-world application. In addition, other study have investigated the dynamic relationship between BIM education and students' career development, assessed the effectiveness of university BIM education and industry recruitment of BIM personnel, and indicated that the two should actively collaborate which created a benchmark for research in BIM education and career development for college students (Wu & Issa, 2014).

However, although all of the above studies have focused on BIM education, learning processes, student employment, and career development, existing research only remains at the level of descriptive statistical analysis. Given the increasingly competitive employment environment for graduates, it is particularly important to study how to enhance students' competitiveness in the workplace.

The objective of this study is to investigate the specific relationship between BIM education in schools and the effectiveness of BIM education. The effectiveness of BIM education in this study is defined as BIM-related career choices and career development behaviors. By combining social cognitive theory and constructing a model, this study conducts an empirical analysis of the effectiveness of BIM technology education in the field of construction management and its influencing factors. By understanding the internal driving mechanisms, targeted guidance and suggestions can be provided for society, universities, and individual students, effectively improving the effectiveness of BIM education in the field of construction management.

4.2 Literature Review

4.2.1 BIM Education

In recent years, BIM education has received extensive attention from scholars worldwide, as well as in China. BIM technology has been gradually incorporated into the higher education of AEC (Architecture, engineering, and construction) related disciplines. There are early studies on BIM-related curriculum system reform, with many countries actively introducing BIM technology into their curriculum (Sacks & Barak, 2010; Sharag-Eldin & Nawari, 2010). There have been studies evaluating BIM teaching methods, finding that students have insufficient knowledge in coordination, interoperability, and clash detection (Kolarić et al., 2019). There are also studies describing students' perspective on the implementation of BIM education in colleges and universities, with feedback from the industry informing curricula adjustments (Adekunle et al., 2022). Swinburne University of Technology carried out a study of AEC majors' perceptions on the practice of BIM, which provided a basis for BIM educators and AEC industry employers offering recommendations on how to optimize resources (Zou et al., 2019). Literature analysis was used to identify relevant BIM competencies and examples of teaching practices, interviews were conducted with industry BIM experts to determine their needs for new hires, and students were surveyed to document their expectations of BIM in the classroom to develop learning objectives (Lucas, 2017). In addition, a study of college students' willingness to learn BIM and influencing factors showed that 98% of students expressed interest in learning BIM, while the reasons for postponing learning were mainly due to a lack of rules and regulations, standard operating procedures, and lack of profitability (Lopez-Zaldivar et al., 2017).

In China, the concept of BIM technology in the curriculum system started late, but the importance of BIM technology has been widely recognized in the education industry (Jin et al., 2018). Numerous scholars have studied the reform of China's BIM education system. For instance, Li, Jingming et al. proposed the cultivation of BIM specialists among China's undergraduate architectural students (Li et al., 2020). Similarly, Xu, Jin et al. have combined new construction engineering technologies with BIM and introduced them into the teaching program (Xu et al., 2018).

However, there are few studies related to BIM education in schools and students' employment and career development from the perspective of individual students. Therefore, it is important to investigate the dynamic relationship between BIM education in universities and students' career development, and thus to assess the effectiveness of BIM education in university and industry recruitment practices (Wu & Issa, 2014). This approach will serve to better understand the internal drivers of the effectiveness of BIM education and the relationship between their influences.

At present, there are still many challenges to BIM education in China's colleges and universities, with the vast majority of research focuses on the reform of the education system, while very little research is oriented to the whole process of BIM education for students in school, and explores the specific impact of the path from BIM education in school to BIM employment and career development. In summary, this study concludes that it is very important to conduct research on BIM education for construction management majors in colleges and universities. From the perspective of students, it is conducive to improving the professional strength of BIM technology and employment competitiveness. From the perspective of colleges and universities, it can continuously improve the effectiveness of BIM education, and promote the long-term development of construction management graduates in the discipline of BIM technology.

4.2.2 Social Cognitive Theory

According to Bandura, individual cognition plays an important role in behavior, and on this basis developed Social Cognitive Theory (SCT). The theory proposes that the individual and the environment dynamically interact with each other. He emphasized the important role of human cognitive factors in behavior and described the formation and maintenance mechanisms of human behavior (Bandura, 1986; Bandura & Walters, 1977).

Among them, individual factors are mainly cognitive factors, which are expressed as self-efficacy. Self-efficacy is the judgment of a person's ability to accomplish a particular task, which is the result of the interaction between the external environment, other self-regulatory mechanisms, and the individual's ability, experience, and performance (Bandura, 1986). According to Bandura, an individual initiates, regulates, and maintains his or her behavior on the basis of self-efficacy (Bandura, 1982; Schunk & Zimmerman, 1996). In addition, Igbaria and Iivari argued that self-efficacy plays an important role in an individual's motivation and behavior, where people with high self-efficacy are more likely to behave accordingly (Igbaria & Iivari, 1995).

Research on social cognitive theory has been applied to the fields of management, sociology, and economics. It has also been applied to learning processes, self-efficacy development, and classroom teaching and learning (Adebusuyi et al., 2022; Stewart et al., 2020). Based on social cognitive theory, many scholars have studied the interrelationships between the environment, individuals, and behaviors, as a means to construct models, formulate hypotheses, and conduct empirical studies (Almulla & Al-Rahmi, 2023; Cai & Shi, 2022; Kursan Milaković, 2021; Zhao & Zhou, 2021). Following that precedent, this study aims to explore the influence of environmental factors and individual cognitive factors on students' career choice behaviors in college BIM education and construct a model of factors influencing the effectiveness of BIM education for construction management majors.

The social cognitive theory emphasizes how individual cognitive factors influence behavior, particularly the crucial role of self-efficacy in motivating and guiding behavioral choices. In the context of BIM education, students' self-efficacy—their belief in their ability to successfully use BIM technology to complete a task which can significantly impact their motivation, involvement, and ultimately their learning outcomes. When students feel confident in their ability to master BIM tools, they are more likely to engage actively in class activities, explore advanced BIM features, and persevere through challenges. Conversely, if they lack confidence in their BIM skills, they may avoid activities that require those skills, which can limit their opportunities for growth and learning. Social cognitive theory provides an effective framework for BIM education. Educators can enhance students' self-efficacy by creating supportive learning environments that promote their learning outcomes. For example, providing hands-on opportunities that enable students to apply BIM technology in real or simulated professional environments can enhance their sense of skill mastery and career readiness. By identifying and adapting environmental factors that affect students' self-efficacy and motivation to learn, BIM education can more effectively promote students' skill development and career readiness.

Thus, applying social cognitive theory to the study of BIM education not only deepens our understanding of students' intrinsic learning motivation but also offers a theoretical basis for developing more effective educational strategies and interventions. This theoretical framework inspires researchers and educators to explore ways to improve the effectiveness of BIM education by enhancing students' self-efficacy and optimizing the learning environment to provide better career preparation for construction management students.

4.3 Materials and Methods

4.3.1 Model Construction and Research Hypothesis

Based on the Social Cognitive Theory, this study argues that students' BIM educational effectiveness is determined by both the individual and the environment. In this study, the behavioral variable is defined as BIM educational effectiveness, where educational effectiveness is the students' Learning Outcome, which is the evaluation and judgment of the changes in knowledge and comprehension (cognition), attitudes and values (affect), practical skills (skills), and behaviors that students have gained through teaching and learning activities or other school experiences (Fatimah et al., 2023). Specifically, students' attention to BIM-related job information and career prospects, active learning of BIM technology, and other behaviors are reflections of the effectiveness of BIM education in school. The Individual refers to students' BIM career self-efficacy, where students' cognition of BIM technology affects the effectiveness of BIM education. Based on previous studies, this study divides the variable BIM career self-efficacy into two secondary variables: BIM career decision self-efficacy and BIM career decision coping efficacy. The former represents the student's self-assessment or confidence in their ability to successfully complete various tasks related to BIM; The latter refers to the career frustration and risk coping ability of students engaged in BIM-related work. The Environment refers to BIM education, where the strengths and weaknesses of BIM education in colleges and universities also affect the effectiveness of BIM education.

Based on this, a total of environmental factors including BIM theory education environment (TE), BIM practice education quality (PQ), cognitive factors including BIM career decision self-efficacy (CDSE), BIM career decision coping efficacy (CDCE), and behavioral factors as BIM education effectiveness (EE), are incorporated in this study. Based on social cognitive theory, the following hypotheses can be proposed, as shown in Table 4.1.

This study draws on the theoretical basis of existing research utilizing Social Cognitive Theory in combination with construction management majors and college BIM education characteristics, in order to construct a model of factors influencing the effectiveness of BIM education for construction management majors, as shown in Fig. 4.1.

4.3.2 Questionnaire Design

This study combines the knowledge frontier of existing research (Lent et al., 2016) and the actual situation of BIM education for construction management majors in China, in order to develop a measurement scale of factors influencing the effectiveness of BIM education for construction management majors. The outcome informs questionnaire design. The questionnaire comprises two parts, the first part surveys demographic details, while the second part proposes scaled questions. A five-point Likert scale is used, with 1–5 indicating "completely disagree", "basically disagree", "neutral", "basically agree" and "completely Agree". Specific scale titles are shown in Exhibit 1.

Table 4.1 Research hypothesis summary

Hypothesis number	Research hypothesis
H1	TE has a significant positive impact on CDSE
H2	TE has a significant positive impact on CDCE
H3	PQ has a significant positive impact on CDCE
H4	PQ has a significant positive impact on CDSE
H5	CDSE has a significant positive impact on EE
H6	CDCE has a significant positive impact on EE
H7	PQ has a significant positive impact on EE
H8	TE has a significant positive impact on EE

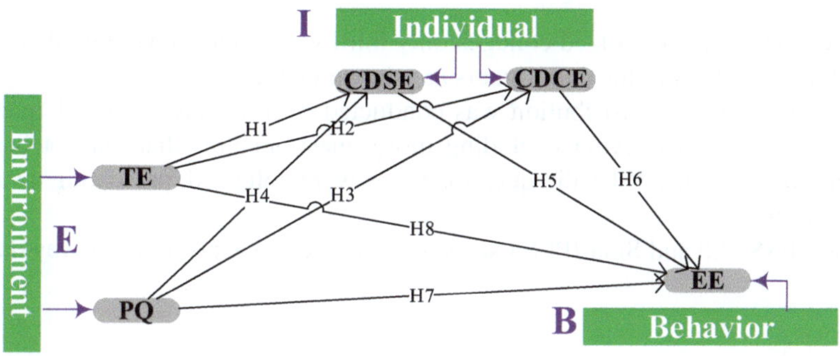

Figure 4.1 Proposed structural model

4.3.3 Sampling and Data Collection

This questionnaire survey was launched in April 2023. Prior to this, the team conducted a pre-survey pilot among the faculty and students of the Chengdu University of Technology majoring in construction management. This pilot informed minor revisions of the questionnaire while testing validity.

Samples were drawn from construction management faculties of universities in the Sichuan and Chongqing regions of China. To ensure non-bias validity of the study, a random sampling method was undertaken. Considering that the core courses of the construction management major are usually studied in the senior year, with behavioral cognition of students in senior year regarding future employment being clearer, only students in their third and fourth years were selected in undergraduate

Table 4.2 The details of the participants

Characteristics		Samples	
		Frequency	Proportion(%)
Gender	Male	168	47.86
	Female	183	52.14
Grade	Junior	256	73.01
	Senior	95	26.99
School type	Application	92	26.21
	Applied research	134	38.18
	Private	125	35.61
The industry area in which students expect to be employed	Architectural engineering	147	41.92
	Engineering construction	76	21.54
	Electromechanical engineering	13	3.64
	Real estate development	67	19.07
	Financial company	30	8.73
	other	18	5.09
The relevance of students' expected careers to BIM technology	Related	248	70.66
	Unrelated	103	29.34

colleges, while in specialized colleges only third year students were invited to participate. The details of the participants are shown in Table 4.2.

The questionnaire distribution was conducted over 12 days and 721 questionnaires were collected. After excluding those questionnaires that did not pass the polygraph questions, 351 valid questionnaires were collected, presenting a validity rate of 48.8%.

The SPSS27.0 and SmartPLS4 software was used to analyze the survey data.

4.4 Results

4.4.1 Reliability and Validity Tests

Reliability refers to the internal consistency between the measured indicators, which can be verified by Cronbach's alpha (≥0.5), composite reliability (CR) (≥0.7), and factor loadings (≥0.5) (Bagozzi & Youjae, 1988; Fornell & Larcker, 1981). The test results of this study showed that all latent variables had Cronbach's alpha greater than 0.5, CR values greater than 0.7, and factor loadings all greater than 0.5. Overall, they passed the reliability test.

Since the 26 items of this study's scale are self-administered scales with no a priori theory except for the BIM Career Decision Self-Efficacy (CDSE) and the BIM Career Decision Coping Self-Efficacy (CDCE), an Exploratory Factor Analysis (EFA) is required. EFA is a process of exploring for factors based on the unknown influences, derived from the data. Therefore, the main purpose of EFA is to find the number of factors that influence the observed variables and the degree of correlation between each factor and each observed variable (Schreiber, 2021; Taherdoost et al., 2014).

This study used SPSS 27.0 to reduce the dimensionality of the 26 BIM technology education effectiveness influencing factor items. During the exploratory factor analysis, the maximum variance method was used to rotate the factors so that the factor loadings varied to greater or lesser extremes, and those items with maximum rotated loadings of less than 0.5 were deleted (Saeed et al., 2022). Therefore, question 21, 'the school's BIM technology course design and skills competition is highly relevant to my future job', was deleted.

The final results show that there are five public factors with eigenvalues greater than 1, and the total explained variance is 78.452%. A total of five-dimensional public factors were extracted. These are: BIM theoretical education environment, BIM practical education quality, BIM career decision self-efficacy, BIM career decision coping efficacy, and BIM technical education effectiveness. This result accords with the design of this study.

Convergence validity was assessed by Average Variance Extracted (AVE), where AVE greater than 0.5 indicates that the structure has good convergence (Joe F. Hair et al., 2012). In this study, the AVE values were all greater than 0.5, thus the convergent validity was good. The square root of AVE for each variable was higher than the correlation coefficient of the latent variable with other latent variables (Fornell & Larcker, 1981), indicating that this study has good discriminant validity. In conclusion, reliability and validity were confirmed. Reliability and validity test indices are shown in Tables 4.3 and 4.4.

4.4.2 Model Fitting Results

In this study, the explanatory power of the structural model was further analyzed using SmartPLS 4.0 to test the validity of the structural equation model, including the coefficient of determination R2, standardized root-mean-square residuals (SRMR), predictive correlation coefficient Q2, and Goodness-of-Fit Indices (GoF) of the overall fit of the model. In general, R2 measures the explanatory power of the model, and a higher value of R2 represents a higher degree of model explanation. R2 > 0.1, and thus the explanatory power of the model is acceptable (Peng et al., 2020). According to the criterion of Hu and Bentler (1998), SRMR<0.1 is also acceptable. The predictive correlation coefficient, Q2 (Stone, 1974), provides good predictive relevance where Q2 > 0 (Cohen, 1977). All the metrics in this study passed this test (Sarstedt et al., 2014). Table 4.5 details the parameters of each indicator for model fit.

Table 4.3 Reliability and validity test results

Variable	Topic item	Factor loading	Cronbach's alpha	CR	AVE
TE	TE1	0.718	0.928	0.938	0.717
	TE2	0.871			
	TE3	0.797			
	TE4	0.805			
	TE5	0.932			
	TE6	0.935			
PQ	PQ1	0.893	0.910	0.914	0.780
	PQ2	0.934			
	PQ3	0.819			
CDSE	SE1	0.818	0.949	0.949	0.701
	SE2	0.835			
	SE3	0.852			
	SE4	0.809			
	SE5	0.898			
	SE6	0.800			
	SE7	0.808			
	SE8	0.873			
CDCE	CE1	0.805	0.904	0.905	0.704
	CE2	0.839			
	CE3	0.851			
	CE4	0.860			
EE	EE1	0.875	0.927	0.927	0.760
	EE2	0.851			
	EE3	0.900			
	EE4	0.861			

Table 4.4 Discriminant results

	TE	PQ	CDSE	CDCE	EE
TE	0.847				
PQ	0.675***	0.883			
CDSE	0.620***	0.600***	0.837		
CDCE	0.446***	0.572***	0.610***	0.839	
EE	0.499***	0.469***	0.830***	0.605***	0.872

Note: ***, $p < 0.001$; **, $p < 0.01$; Diagonal line is SQRT(AVE)

4.4.3 Hypothesis Testing

The fitting results of the model are shown in Fig. 4.2. It can be seen from Fig. 4.2 that the significance of all paths was less than 0.05, except that H7 and H8 was greater than 0.05, and the standardized regression coefficients were all positive except H7, so the original hypotheses1-hypotheses 6 are accepted, the original hypothesis7 and hypothesis8 is rejected. Among them, the BIM education effectiveness was influenced by BIM career decision self-efficacy and BIM career decision coping efficacy, and the order of influence was the BIM career decision self-efficacy(0.707) > the BIM career decision coping efficacy(0.182); the BIM career decision self-efficacy was influenced by BIM theory education environment and BIM practice education quality, and the order of influence was BIM theory education environment(0.403) > BIM practice education quality (0.324); the BIM career decision coping efficacy was influenced by the order BIM practice education quality (0.452) > BIM theory education environment (0.124) (Table 4.6)

To further validate the internal driving mechanism of the factors influencing the effectiveness of BIM education for construction management majors, this study used the Bootstrap method to analyze the influence of mediating effects under different paths. The characteristic indirect effects are shown in Table 4.7 the total effects are shown in the appendix.

4.5 Discussion

This study based on the Social Cognitive Theory to explore the factors that influence the effectiveness of BIM education in construction management students. The relationship between BIM theory education environment, BIM practice education quality, BIM career decision self-efficacy, BIM career decision coping efficacy and BIM education effectiveness were examined.

4.5.1 BIM Theory Education Environment Dimension

As our research found that the theoretical BIM education environment of construction management majors has a very significant positive indirect effect on the effectiveness of BIM education through BIM career decision self-efficacy ($\beta = 0.285$, $p < 0.001$), in which BIM career decision self-efficacy plays a mediating role. The student group can form clearer BIM cognition and develop higher BIM career self-efficacy. This occurs through the education of BIM-related theoretical courses in school and the incentives provided related to BIM education (e.g., BIM labs, BIM school-enterprise cooperative internships, etc.). That is to say, they have the ability and confidence needed to engage in BIM-related industries, and can make sound career decisions. Research on the employment quality of college students shows that learning experience has a significant impact on employment quality (Aamodt & Havnes, 2008), of which course learning has the greatest impact (Ehiyazaryan & Barraclough, 2009).

4.5.2 BIM Practice Education Quality Dimension

We can see from the results that BIM practice education quality had a highly significant positive indirect effect on BIM education effectiveness through BIM career decision self-efficacy ($\beta = 0.229$, $p < 0.001$), where BIM career decision self-efficacy plays a fully mediating role. On the other hand, there was a more significant positive indirect effect ($\beta = 0.082$, $p < 0.01$) on BIM educational effectiveness through BIM career decision-coping efficacy. Here, BIM career decision-coping efficacy played a partially mediating role. Through BIM practical educational activities, such as course design and skill competitions, school students have a certain

Table 4.5 Model fitting index parameters

Latent variable	R^2(>0.1)	SRMR(<0.1)	Q^2(>0)	AVE	GoF
CDCE	0.289	0.056	0.219	0.777	0.59
CDSE	0.435		0.315	0.737	
TE	–		–	0.767	
EE	0.620		0.506	0.820	
PQ	–			0.847	
Average	0.448		–	0.790	

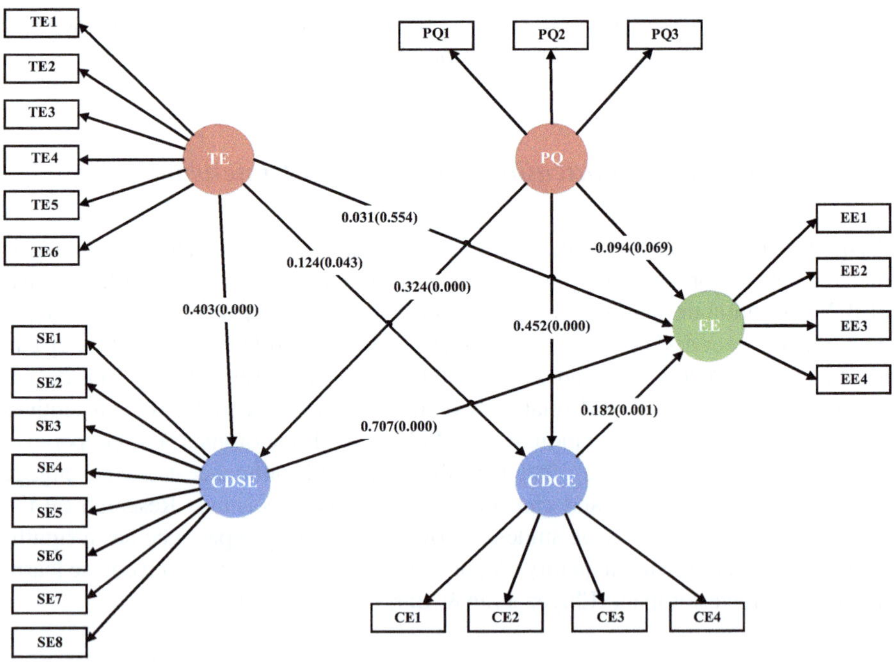

Figure 4.2 Structural equation modeling analysis diagram

Table 4.6 Results of hypothesis testing

Hypothesis	Path	Original sample(0)	Standard deviation(STDEV)	T statistics(|0/STDEV|)	P values
H1	TE -> CDSE	0.403	0.053	7.544	0.000(***)
H2	TE -> CDCE	0.124	0.061	2.024	0.043(*)
H3	PQ -> CDCE	0.452	0.061	7.465	0.000(***)
H4	PQ -> CDSE	0.324	0.057	5.727	0.000(***)
H5	CDSE -> EE	0.707	0.053	13.331	0.000(***)
H6	CDCE -> EE	0.182	0.053	3.404	0.001(***)
H7	PQ -> EE	−0.094	0.051	1.822	0.069
H8	TE -> EE	0.031	0.052	0.592	0.554

BIM career self-efficacy, a certain knowledge of BIM-related workings, and the ability to judge risks, which positively affects final BIM education effectiveness. Many scholars have pointed out the importance of practical activities in improving students' employment outcomes, and the need to continuously strengthen students' practical ability (Lu, 2021). Thus, practical activities such as skill competitions can have an extremely positive impact on college students learning and employability (Liu et al., 2018).

4.5.3 BIM Career Self-Efficacy Dimension

As our research found that the original hypotheses 5 and hypotheses 6 were accepted, the effectiveness of BIM education was influenced by the BIM career decision self-efficacy and BIM career decision coping efficacy. The concept of career self-efficacy was first proposed by Hackett and Betz as a belief in an individual's ability to engage in a particular career (Hackett & Betz, 1981). Lent further subdivided BIM career self-efficacy into career decision self-efficacy and career decision coping efficacy. Career decision self-efficacy focuses on confidence in exploring career options, while career decision coping efficacy is the belief in being able to manage decision barriers or failures (Lent et al., 2016). Construction management students' BIM career decision self-efficacy had a highly significant positive effect on BIM educational effectiveness ($\beta = 0.707$, $p < 0.001$), while BIM career decision coping efficacy likewise had a highly significant positive effect ($\beta = 0.182$, $p < 0.001$). The higher students' BIM career decision self-efficacy and coping efficacy indicate that they have a clearer perception of their career choices and are better able to cope with risks. This is conducive to students' career development, which in turn has a significant positive impact on the effectiveness of BIM education. Our findings are similar to those of many studies have also shown that career self-efficacy is significantly and positively correlated with the range of career choices (Feiyue, 2002). The higher the career self-efficacy, the more accurate the prediction of career choice behavior. Individuals with high career self-efficacy also perform better in their BIM education

Table 4.7 Structural equation modeling specific indirect effects

Path	Original sample(0)	Standard deviation(STDEV)	T statistics(|0/STDEV|)	P values
PQ -> CDSE -> EE	0.229	0.045	5.067	0.000(***)
PQ -> CDCE -> EE	0.082	0.026	3.149	0.002(**)
TE -> CDCE -> EE	0.023	0.013	1.711	0.087
TE -> CDSE -> EE	0.285	0.042	6.823	0.000(***)

(Gledson & Dawson, 2017). The higher students' BIM career decision self-efficacy and coping efficacy indicate that they have a clearer perception of their career choices and are better able to cope with risks. This is conducive to students' career development, which in turn has a significant positive impact on the effectiveness of BIM education.

4.5.4 Comprehensive Discussion

Some research have shown that college students' studies at school and factors such as institutional support and high-impact practical activities have a positive impact on college students' career identity, while also helping students form clear career perceptions (Kilgo et al., 2015). Subjective perceived educational gains during college have a significant positive impact on students' employment and career development, with knowledge gains having the greatest impact on employment (Vermeulen & Schmidt, 2008). All of the above studies indicate that both theoretical courses and practical education received in school have a positive impact on the employment and career development behavior of college students.

Comparing the four paths of influence of in-school BIM technical education on BIM education effectiveness through BIM career self-efficacy, this study found that three paths of influence were significant, with the exception being the path of influence of BIM theoretical education environment on BIM education effectiveness through BIM career decision-making coping efficacy. I think there's an interesting point about this path that does not have a significant impact. In my opinion, BIM theoretical education environment can provide students with in-depth learning of professional knowledge, which makes their cognition of major and career clearer. That means the career decision-making self-efficacy is higher. At the same time, it is clear that only BIM theoretical education environment has no significant positive impact on students' risk coping ability and decision-making ability in the future workplace, that is, the BIM career decision-making coping efficiency is weak, so the impact on the effectiveness of BIM technology education is not significant. Of the influences of the BIM theoretical education environment on BIM career self-efficacy, the influences of the BIM career decision-making self-efficacy are greater, while the quality of BIM practical education has a strong effect on BIM career decision-making coping efficacy. This may be due to the fact that students' BIM career decision self-efficacy is high because they have a clearer perception of their profession through receiving theoretical coursework and the incentives provided by the school. This gives them a certain level of understanding and confidence in the competencies required for the career they will be pursuing in the future. Many scholars have studied the relationship between school learning, career decision self-efficacy, and career choice behavior which have yielded the resemble results with our research. These studies on various aspects of theoretical education confirm the findings of our study and draw conclusions similar to ours. As our study found, the BIM theoretical education environment can influence the effectiveness of BIM education by influencing BIM career self-efficacy. Many studies have also found the impact of educational environment on career decision self-efficacy. For example, Choi and Kim explored the relationship between motivation, academic achievement, and career decision self-efficacy of undergraduate students majoring in tourism management and hospitality, and found that student's academic performance such as grades significantly affected career decision self-efficacy. At the same time, career decision self-efficacy affected the implementation of career preparation behaviors (Choi & Kim, 2013). College students' academic majors and intrinsic work values significantly influence career decision self-efficacy (Doo & Park, 2019). Research has also shown that teaching and learning situations also affect students' career decision self-efficacy (Plakhotnik et al., 2020). For example, positive feedback in learning improves students' career decision self-efficacy and job search intentions (Deer et al., 2018).

BIM practical education, on the other hand, focuses more on the students' hands-on experience, which is a kind of simulation of the actual future work situation. As a result of participating in these practical activities, they accrue experience and possess certain professional qualities, and moreover become more comfortable when facing possible career setbacks. Our study finds that BIM practical education has a significant impact on students' career decision-making coping ability. We've found some similar results in other studies. These studies provide strong support for the findings of our study. Studies have shown that extracurricular practical activities can effectively improve students' developmental soft power and employment competitiveness (Aliu & Aigbavboa, 2023). In addition, many studies have shown that although the transition of graduates from college to work is just a normal life event (Schlossberg, 2011), a large percentage of graduates experience difficulties in career decision-making (Schoon & Silbereisen, 2009), with the first job delivering undue stress (Dietrich et al., 2012; Wendlandt & Rochlen, 2008). This process is accompanied by a lot anxiety and indecision about career choices (Anderson et al., 2012). The main reason for this is the poor career adaptability of graduates, which Savickas (Savickas, 1997) argues is centered on poor career decision-making coping skills needed to effectively cope with various career-related challenges.

Although the direct effects of the BIM theoretical education environment and the quality of BIM practical education covered by the environmental factor of BIM education were not statistically significant, the path coefficients were small. Simply, there was only a very weak effect. However, after the mediation of BIM career self-efficacy, the total effects of theoretical courses and incentives and practical activities such as classroom settings and skill competitions on BIM education effectiveness are significant. Interestingly, of these, the quality of BIM practical education has a negative and weak effect on the effectiveness of BIM education. This is contrary to the findings of existing studies, and is probably due to respondents not having a positive experience with current BIM-related practical activities. This study also found that the BIM theoretical education environment and the quality of BIM practical education had a greater influence on the pathway of BIM education effectiveness through BIM career self-efficacy, than the pathway of BIM career decision-making coping efficacy alone on the pathway of BIM education effectiveness. These are strong indications that the environment influences behavior more indirectly through cognition. Therefore, BIM education for construction management majors should focus on improving students' career self-efficacy, and studies have shown that career self-efficacy can be improved by providing more career-related information, along with courses that offer specific professional experience (Aliu & Aigbavboa, 2023). Colleges and universities should enhance students' career self-efficacy by optimizing the BIM education environment in schools, and thereby significantly improving the effectiveness of BIM education.

By comparing the impact of both types of BIM career self-efficacy on the effectiveness of BIM education, it is obviously to reveal that the impact of BIM career decision self-efficacy is greater in the student population. This may be due to the lower impact of career decision-making coping efficacy, as school students have not yet had work experience and are thus less able to cope with disappointment or resort to alternatives when preference barriers are encountered. It has also been confirmed that students with work experience significantly outperform inexperienced students in making plans and selecting goals [50]. In addition, an analysis of all the pathways affecting the effectiveness of BIM technical education revealed that the highest pathway coefficient was BIM career decision-making self-efficacy ($\beta = 0.707$, $p < 0.001$), indicating that BIM career decision-making self-efficacy had the greatest impact on final BIM education effectiveness.

Established studies have also shown that career decision self-efficacy is positively correlated with career inclination and career optimism [51]. There is a strong correlation between career decision self-efficacy and career adaptability [52], while there is a negative correlation between career self-efficacy and career indecisiveness [53]. Consistent with our findings, existing studies fully illustrate that higher occupational self-efficacy results in clearer career inclinations, more optimistic career attitudes, more decisiveness in the face of career choices, and less hesitation and indecisiveness, all of which are conducive to making clear career choices. That is to say, occupational self-efficacy has a positive and significant influence on final career choice. Therefore, we believe that improving students' career self-efficacy has a profound impact on enhancing the effectiveness of BIM education.

4.6 Conclusion

Utilizing the findings of both Chinese and non-Chinese literature, this study examined the influence mechanism of BIM education on BIM education effectiveness in construction management majors. It did so using a constructed model of influencing factors based on social cognitive theory, with conclusions drawn through empirical analysis. The results of the study show that the direct influence of in-school BIM education on BIM education effectiveness is not significant. However, considering the mediating variable of BIM career self-efficacy, there is a significant positive indirect influence on BIM education effectiveness.

In alignment with this finding, and in order to enhance the effectiveness of BIM education for construction management majors in colleges and universities, this study proposes the following measures:

1. Optimize the theoretical curriculum and increase practical cases so that students can better understand the practical application of BIM technology, thus enhancing their sense of BIM career self-efficacy.
2. Strengthen the organization and guidance of practical activities, encourage students to actively participate in practical activities such as course design and skill competitions, and improve their career quality and adaptability.
3. Provide students with career-related consulting and counseling services to help them make wise career decisions.

Through the above measures, students' BIM career self-efficacy can be improved, thus further enhancing the effect of BIM education in universities, while laying a solid foundation for students' career development and employment quality. This can be expected to improve BIM education of engineering management majors, meet the industry's demand for BIM skills, and promote the sustainable development of the engineering industry.

References

Aamodt, P. O., & Havnes, A. (2008). Factors affecting professional job mastery: Quality of study or work experience? *Quality in Higher Education, 14*(3), 233–248.

Adebusuyi, A. S., Adebusuyi, O. F., & Kolade, O. (2022). Development and validation of sources of entrepreneurial self-efficacy and outcome expectations: A social cognitive career theory perspective. *The International Journal of Management Education, 20*(2), 100572.

Adekunle, S. A., Aigbavboa, C. O., & Ejohwomu, O. A. (2022). Understanding the BIM actor role: A study of employer and employee preference and availability in the construction industry. *Engineering, Construction and Architectural Management*. https://doi.org/10.1108/ECAM-08-2021-0714. https://www.emerald.com/insight/content/doi/10.1108/ECAM-08-2021-0714/full/html

Aliu, J., & Aigbavboa, C. (2023). Reviewing the roles of extracurricular activities in developing employability skills: A bibliometric review. *International Journal of Construction Management, 23*(10), 1623–1632. https://doi.org/10.1080/15623599.2021.1995807

Almulla, M. A., & Al-Rahmi, W. M. (2023). Integrated social cognitive theory with learning input factors: The effects of problem-solving skills and critical thinking skills on learning performance sustainability. *Sustainability, 15*(5), 3978.

Anderson, M. L., Goodman, J., & Schlossberg, N. K. (2012). *Counseling adults in transition: Linking Schlossberg's theory with practice in a diverse world* (4th ed.). Springer Publishing Company.

Ao, Y., Peng, P., Li, J., Li, M., Bahmani, H., & Wang, T. (2022). What determines BIM competition results of undergraduate students in the architecture, engineering and construction industry? *Behavioral Sciences, 12*(10). https://doi.org/10.3390/bs12100360

Bagozzi, R. P., & Youjae, Y. (1988). On the evaluation of structural equation models. *Journal of the Academy of Marketing Science, 16*(1), 74–94. https://doi.org/10.1177/009207038801600107. https://journals.sagepub.com/doi/abs/10.1177/009207038801600107

Bandura, A. (1982). Self-efficacy mechanism in human agency. *American Psychologist, 37*(2), 122.

Bandura, A. (1986). *Social foundations of thought and action: A social cognitive theory*.(pp. xiii, 617-xiii, 617). Prentice-Hall, Inc.

Bandura, A., & Walters, R. H. (1977). *Social learning theory* (Vol. 1). Englewood cliffs Prentice Hall.

Cai, Y., & Shi, W. (2022). The influence of the community climate on users' knowledge-sharing intention: The social cognitive theory perspective. *Behaviour & Information Technology, 41*(2), 307–323.

Castro-Lacouture, D., Irizarry, J., & Ashuri, B. (2014). Construction research congress 2014 construction in a global network. In *Proceedings of the 2014 construction research congress*.

Choi, K., & Kim, D.-Y. (2013). A cross cultural study of antecedents on career preparation behavior: Learning motivation, academic achievement, and career decision self-efficacy. *Journal of Hospitality, Leisure, Sport & Tourism Education, 13*, 19–32. https://doi.org/10.1016/j.jhlste.2013.04.001. https://www.sciencedirect.com/science/article/pii/S1473837613000105

Cohen, J. (1977). Chapter 1: The concepts of power analysis. In J. Cohen (Ed.), *Statistical power analysis for the behavioral sciences* (pp. 1–17). Academic Press. https://www.sciencedirect.com/science/article/pii/B9780121790608500062

Deer, L. K., Gohn, K., & Kanaya, T. (2018). Anxiety and self-efficacy as sequential mediators in US college students' career preparation. *Education + Training, 60*(2), 185–197. https://doi.org/10.1108/ET-07-2017-0096

Dietrich, J., Jokisaari, M., & Nurmi, J.-E. (2012). Work-related goal appraisals and stress during the transition from education to work. *Journal of Vocational Behavior, 80*(1), 82–92. https://doi.org/10.1016/j.jvb.2011.07.004. https://www.sciencedirect.com/science/article/pii/S000187911100100X

Doo, M. Y., & Park, S. H. (2019). Effects of work value orientation and academic major satisfaction on career decision-making self-efficacy. *Higher Education, Skills and Work-Based Learning, 9*(4), 550–562. https://doi.org/10.1108/HESWBL-09-2018-0088

Ehiyazaryan, E., & Barraclough, N. (2009). Enhancing employability: Integrating real world experience in the curriculum. *Education+ Training, 51*(4), 292–308.

Fatimah, F. S., Asy'ari, H., Sandria, A., & Nasucha, J. A. (2023). Learning Fiqh based on the TAPPS (think aloud pair problem solving) method in improving student learning outcomes. *At-Tadzkir: Islamic. Education Journal, 2*(1), 1–15.

Feiyue, J. (2002). *The career self-efficacy theory and its'application to occupational choices of senior university students* (Master). Nanjing Normal University, N1 – Guo Benyu.

Fornell, C., & Larcker, D. F. (1981). Evaluating structural equation models with unobservable variables and measurement error. *Journal of Marketing Research, 18*(1), 39–50. https://doi.org/10.1177/002224378101800104

Gledson, B. J., & Dawson, S. (2017). Use of simulation through BIM-enabled virtual projects to enhance learning and soft employability skills in architectural technology education. In *Building information modelling, building performance, design and smart construction* (pp. 79–92). Springer International Publishing.

Gu, S. (2004). China's industrial technology: Market reform and organisational change.

Hackett, G., & Betz, N. E. (1981). A self-efficacy approach to the career development of women. *Journal of Vocational Behavior, 18*(3), 326–339. https://doi.org/10.1016/0001-8791(81)90019-1. https://www.sciencedirect.com/science/article/pii/0001879181900191

Hair, J. F., Sarstedt, M., Ringle, C. M., & Mena, J. A. (2012). An assessment of the use of partial least squares structural equation modeling in marketing research. *Journal of the Academy of Marketing Science, 40*(3), 414–433. https://doi.org/10.1007/s11747-011-0261-6

Hu, L.-T., & Bentler, P. M. (1998). Fit indices in covariance structure modeling: Sensitivity to underparameterized model misspecification. *Psychological Methods, 3*, 424–453. https://doi.org/10.1037/1082-989X.3.4.424

Igbaria, M., & Iivari, J. (1995). The effects of self-efficacy on computer usage. *Omega, 23*(6), 587–605.

Jin, R., Yang, T., Piroozfar, P., Kang, B.-G., Wanatowski, D., Hancock, C. M., & Tang, L. (2018). Project-based pedagogy in interdisciplinary building design adopting BIM. *Engineering, Construction and Architectural Management, 25*(10), 1376–1397.

Kilgo, C. A., Ezell Sheets, J. K., & Pascarella, E. T. (2015). The link between high-impact practices and student learning: Some longitudinal evidence. *Higher Education, 69*, 509–525.

Kolarić, S., Mandičák, T., Vukomanović, M., & Mesároš, P. (2019). The influence of BIM education improvement on raising BIM awareness in Croatia and Slovakia. *Periodica Polytechnica: Architecture, 50*(1), 21–29. https://search.ebscohost.com/login.aspx?direct=true&db=edb&AN=136688625&lang=zh-cn&site=eds-live

Kursan Milaković, I. (2021). Purchase experience during the COVID-19 pandemic and social cognitive theory: The relevance of consumer vulnerability, resilience, and adaptability for purchase satisfaction and repurchase. *International Journal of Consumer Studies, 45*(6), 1425–1442.

Lent, R. W., Ezeofor, I., Morrison, M. A., Penn, L. T., & Ireland, G. W. (2016). Applying the social cognitive model of career self-management to career exploration and decision-making. *Journal of Vocational Behavior, 93*, 47–57. https://doi.org/10.1016/j.jvb.2015.12.007. https://www.sciencedirect.com/science/article/pii/S0001879115300245

Li, L. (2018). China's manufacturing locus in 2025: With a comparison of "Made-in-China 2025" and "Industry 4.0". *Technological Forecasting and Social Change, 135*, 66–74.

Li, J., Afsari, K., Li, N., Peng, J., Wu, Z., & Cui, H. (2020). A review for presenting building information modeling education and research in China. *Journal of Cleaner Production, 259*, 120885.

Liu, L., Hu, F., Jiang, Z., & Xu, H. (2018). Study on training mode of promoting undergraduates' innovation and entrepreneurial ability with subject competition. In: *2018 3rd international conference on education, sports, arts and management engineering (ICESAME 2018)*.

Lopez-Zaldivar, O., Verdu-Vazquez, A., Gil-Lopez, T., & Lozano-Diez, R. V. (2017). The implementation of building information modeling technology in university teaching: The case of the Polytechnic University of Madrid. *International Journal of Engineering Education, 33*(2), 712–722.

Lu, H.-F. (2021). Enhancing university student employability through practical experiential learning in the sport industry: An industry-academia cooperation case from Taiwan. *Journal of Hospitality, Leisure, Sport & Tourism Education, 28*, 100301.

Lucas, J. D. (2017). Identifying learning objectives by seeking a balance between student and industry expectations for technology exposure in construction education. *Journal of Professional Issues in Engineering Education and Practice, 143*(3). https://doi.org/10.1061/(ASCE)EI.1943-5541.0000318

Ma, J., & Tao, Y. (2023). Learning outcomes of civil engineering students in PBL based on building information modeling. *International Journal of Emerging Technologies in Learning (Online), 18*(7), 89.

Martek, I., Wu, W., Oraee, M., & Hosseini, M. R. (2021). Educating the "T-shaped" BIM professional: Lessons from academia. In *BIM teaching and learning handbook* (pp. 247–261). Routledge.

Peng, L., Tan, J., Deng, W., & Liu, Y. (2020). Farmers' participation in community-based disaster management: The role of trust, place attachment and self-efficacy. *International Journal of Disaster Risk Reduction, 51*, 101895. https://doi.org/10.1016/j.ijdrr.2020.101895. https://www.sciencedirect.com/science/article/pii/S2212420920313972

Peng, P., Ao, Y., Li, M., Wang, Y., Wang, T., & Bahmani, H. (2022). Building information modeling learning behavior of AEC undergraduate students in China. *Behavioral Sciences, 12*(8). https://doi.org/10.3390/bs12080269

Plakhotnik, M. S., Krylova, A. V., & Maslikova, A. D. (2020). Does participation in case competitions improve career decision-making self-efficacy of university students? *Education + Training, 62*(6), 659–675. https://doi.org/10.1108/et-01-2020-0021

Sacks, R., & Barak, R. (2010). Teaching building information modeling as an integral part of freshman year civil engineering education. *Journal of Professional Issues in Engineering Education and Practice, 136*(1), 30–38. https://doi.org/10.1061/(ASCE)EI.1943-5541.000000

Saeed, B., Tasmin, R., Mahmood, A., & Hafeez, A. (2022). Development of a multi-item operational excellence scale: Exploratory and confirmatory factor analysis. *The TQM Journal, 34*(3), 576–602.

Sarstedt, M., Ringle, C. M., Smith, D., Reams, R., & Hair, J. F. (2014). Partial least squares structural equation modeling (PLS-SEM): A useful tool for family business researchers. *Journal of Family Business Strategy, 5*(1), 105–115. https://doi.org/10.1016/j.jfbs.2014.01.002. https://www.sciencedirect.com/science/article/pii/S1877858514000060

Savickas, M. L. (1997). Career adaptability: An integrative construct for life-span, life-space theory. *The Career Development Quarterly, 45*(3), 247–259. https://doi.org/10.1002/j.2161-0045.1997.tb00469.x

Schlossberg, N. K. (2011). The challenge of change: The transition model and its applications. *Journal of Employment Counseling, 48*(4), 159–162. https://doi.org/10.1002/j.2161-1920.2011.tb01102.x

Schoon, I., & Silbereisen, R. K. (2009). Conceptualising school-to-work transitions in context. In *Transitions from school to work: Globalization, individualization, and patterns of diversity* (pp. 3–29). Cambridge University Press.

Schreiber, J. B. (2021). Issues and recommendations for exploratory factor analysis and principal component analysis. *Research in Social and Administrative Pharmacy, 17*(5), 1004–1011.

Schunk, D. H., & Zimmerman, B. J. (1996). Modeling and self-efficacy influences on children's development of self-regulation. In *Social motivation: Understanding children's school adjustment* (pp. 154–180). Cambridge University Press.

Sharag-Eldin, A., & Nawari, N. O. (2010). BIM in AEC education. In *Structures Congress* (pp. 1676–1688). https://doi.org/10.1061/41130(369)153

Shuchi, S., Chowdhury, M., Martek, I., & Hosseini, M. R. (2020). *BIM education at Australian universities*.

Stewart, J., Henderson, R., Michaluk, L., Deshler, J., Fuller, E., & Rambo-Hernandez, K. (2020). Using the social cognitive theory framework to chart gender differences in the developmental trajectory of STEM self-efficacy in science and engineering students. *Journal of Science Education and Technology, 29*, 758–773.

Stone, M. (1974). Cross-Validatory choice and assessment of statistical predictions. *Journal of the Royal Statistical Society: Series B (Methodological), 36*(2), 111–133. https://doi.org/10.1111/j.2517-6161.1974.tb00994.x

Taherdoost, H., Sahibuddin, S., & Jalaliyoon, N. (2014). *Exploratory factor analysis: Concepts and theory*. Post-Print.

Vermeulen, L., & Schmidt, H. G. (2008). Learning environment, learning process, academic outcomes and career success of university graduates. *Studies in Higher Education, 33*(4), 431–451.

Wendlandt, N. M., & Rochlen, A. B. (2008). Addressing the college-to-work transition: Implications for university career counselors. *Journal of Career Development, 35*(2), 151–165. https://doi.org/10.1177/0894845308325646

Wu, W., & Issa, R. R. A. (2014). BIM education and recruiting: Survey-based comparative analysis of issues, perceptions, and collaboration opportunities. *Journal of Professional Issues in Engineering Education and Practice, 140*(2). https://doi.org/10.1061/(ASCE)EI.1943-5541.0000186

Wu, P., Jin, R., Xu, Y., Lin, F., Dong, Y., & Pan, Z. (2021). The analysis of barriers to BIM implementation for industrialized building construction: A China study. *Journal of Civil Engineering and Management, 27*(1), 1–13.

Xu, J., Li, B.-K., & Luo, S.-M. (2018). Practice and exploration on teaching reform of engineering project management course in universities based on BIM simulation technology. *Eurasia Journal of Mathematics, Science and Technology Education, 14*(5), 1827–1835.

Zhao, H., & Zhou, Q. (2021). Socially responsible human resource management and hotel employee organizational citizenship behavior for the environment: A social cognitive perspective. *International Journal of Hospitality Management, 95*, 102749.

Zou, P. X. W., Xu, X., Jin, R., Painting, N., & Li, B. (2019). AEC students' perceptions of BIM practice at Swinburne University of Technology. *Journal of Professional Issues in Engineering Education And Practice, 145*(3), 05019002. https://doi.org/10.1061/(ASCE)EI.1943-5541.0000410

Chapter 5
Factors Driving BIM Learning Performance: Research on China's Sixth National BIM Graduation Design Innovation Competition of Colleges and Universities

Abstract With the popularization and rise in BIM technology usage, BIM education for undergraduate students in architecture, engineering, and construction (AEC) related disciplines has emerged as a priority. This study assesses the BIM learning outcomes of students participating in the National BIM Graduation Design Innovation Competition of Colleges and Universities. In total, 2777 valid questionnaire responses were obtained for this study. The Cronbach's alpha coefficient method and principal component factor analysis method were used to verify the reliability of the data set (Cronbach's alpha = 0.962, KMO = 0.965). The t-test (ANOVA) was used to verify that gender, school type, major, grade, study duration and use BIM related software, as well as other demographic attributes, displayed significant inter-group differences. Seven common factors affecting BIM learning performance were obtained by exploratory factor analysis: (1) ability of the instructor, (2) school (college) atmosphere, (3) teamwork, (4) individual ability, (5) understanding of BIM industry applications, (6) social environment incentives, and (7) achievement demand. Finally, the results of an ordered logistic regression revealed that the demographic attributes of participants, the comprehensive ability of the instructor, teamwork, individual ability, and achievement demand significantly affects BIM learning performance. Based on these findings, this paper puts forward suggestions for improving BIM learning performance and provides theoretical support for BIM education and learning in AEC related undergraduate majors.

Keywords BIM learning performance · BIM education · BIM competition · Undergraduate education · AEC · China

5.1 Introduction

As a major digital technology used by the AEC industry (Ma et al., 2020), building information modeling (BIM) has emerged as the main focus of professional learning in AEC-related higher education (Wang et al., 2020). BIM learning has become

Y. Ao et al., *Building Information Modeling (BIM) Adoption*, Resilient and Friendly Living Environments, https://doi.org/10.1007/978-981-95-3993-2_5

the foundation for the sustainable development of the AEC industry. BIM learning supports the sustainable application of BIM technology across the stages of design, procurement, manufacturing, construction, operation, and maintenance (Yongliang Deng et al., 2020). Although scholars have made great progress in BIM research, with BIM technology now even widely used in Financial Investment Projects (Zhou et al., 2020), nevertheless, there remains a disconnection between theoretical education and engineering practice in BIM learning (AboWardah, 2020).

College and university discipline competitions are based on classroom teaching where competition is used to stimulate student's abilities, such as combining theory with practice, independent thinking and teamwork (Chen Suyue, 2020). Discipline competition can cultivate students' practical innovation ability and promote the all-round development of college students, which has become an important part of practical teaching (Lv et al., 2020). Discipline competition is not only an important means of talent training in universities, but also an important basis by which employers select graduates (Lu et al., 2018). Thus, college students respond by showing great interests in discipline competitions. At present, there are a large number of BIM related discipline competitions being held in China, as well as in other countries. These include: the Annual Brilliance Competition for Infrastructure, the Excellence in BIM Competition for the Global Engineering and Construction Industry, the Hong Kong International BIM Competition, the Building SMART International Competition, the China National Competitions, the National College BIM Graduation Design Innovation Competition, the China Construction Engineering BIM Competition, the Transportation BIM Engineering Innovation Award, the Innovation Cup Building Information Model Application Design Competition, the You-lu Cup National BIM Technology Competition, the Municipal Cup BIM Application Skills Competition, among others. Moreover, BIM-related competitions are sponsored by enterprises and industry associations across various provinces and cities in China. The organizers of discipline competitions usually partner with enterprises. Students not only enhance their BIM skills as demanded by the requirements of the competition, but also leverage their career development after employability (Jie, 2020).

Nationally, across China, graduation design is a compulsory course of AEC-related education, which is the main link between theoretical education and engineering practice. It is formulated to help students smoothly transition from school education to professional practice after graduation (AboWardah, 2020). BIM technology is sits at the core of the future sustainable development of the AEC industry. Therefore, it is crucial to identify the driving factors of BIM learning outcomes and extract those mechanisms that augment BIM learning performance. Consequently, this study investigates the sixth National BIM Graduation Design Innovation Competition of Colleges and Universities (sixth NBGDIC) as a representative vehicle for unpacking the driving factors that accelerate BIM learning. To this end, exploratory factor analysis and ordered logistic regression of returned questionnaire survey results are used.

5.2 Literature Review

5.2.1 BIM and Related Technologies

Building information modeling (BIM) and virtual reality (VR) have attracted increasing attention within the AEC industry over recent years. Virtual reality allows simulation of dangerous, expensive or hard-to-reach environments. It can improve the learning process and motivate students to develop the skills they need to succeed and innovate (Chen et al., 2020). Alizadehsalehi, S., et al. argues that VR technology in the AEC industry makes it easier for students to understand digital data in BIM (Alizadehsalehi et al., 2021). In addition, Wang, L., et al. suggests that BIM can be integrated with other digital technologies or platforms, such as 3D printing, drones, mixed reality and laser scanning (Wang et al., 2020). Digital twins (DT) can use both real-time and historical data to simulate the future state of an asset and predict the consequences of maintenance operations (Coupry et al., 2021). Many researchers try to use BIM as a basis to develop the DT of a building, for either existing buildings or those in the design stage (Cheng et al., 2020). The combination of extended reality (XR) and BIM provides interactive renderings, spatial coordination, and virtual models, and provides many benefits that can improve the performance of AEC projects, including time, cost, quality, and safety. This integration is useful not only during the design and construction phases, but also during maintenance (Cheng et al., 2020). In the AEC industry, XR technology that simulates a building project in a multidimensional digital model and presents multiple aspects of the project can be of great help at all stages of a project (Alizadehsalehi et al., 2020).

5.2.2 Factors Impacting BIM Learning Performance

A systematic examination of existing literature reveals the current, known factors influencing discipline competition outcomes, and these include the demographic attributes of participants, social environment incentives, level of team cooperation, ability of instructor, among others (Hotulainen et al., 2020; Wang Yana & Mao Hanping, 2019). This study develops this knowledge to in an effort to establish the influencing factors impacting the BIM learning performance of AEC undergraduates.

Obviously, the individual idiosyncratic characteristics of participants are the key determinates of learning performance. Previous studies have found that individual ability (thinking, reasoning and task execution) and achievement demand (motivation, attitude or belief) significantly affect students' learning performance (Hotulainen et al., 2020; Niebuhr et al., 1982). Individual thinking mode, reasoning, and task execution ability can affect and assist individuals to deal with the problem of multilateral cooperation between multiple tasks (Liu et al., 2019), thus significantly affecting the final learning outcomes of learners (Fichter et al., 2020). However, individual ability is not the only driving factor of learning performance (Buelow & Cayton, 2020). Peiró, J.M., et al. (2020) found that the learners' perception of the importance of the learning task significantly affects learning motivation, attitude and belief, and thus consequently affects ultimate learning performance (Peiró et al., 2020). Over the duration of a discipline competition, students will draw on their personal store of knowledge and practical ability, yet incentivization through awards and honors can also leverage motivation, which in turn can be expected to impact final learning performance.

In addition to the learner's personal characteristics, the instructors' comprehensive ability will also significantly affect the students' learning performance, good teachers can improve the students' learning scores to varying degrees (Adams, 2005). Some studies have shown that the important factors affecting the learning quality of college students include the professional ability (Eika et al., 2014; Ingvarson, 2019), daily guidance behavior (Escribano et al., 2020), and the effectiveness of guidance (Fleming et al., 2020) of the instructors. Teams participating in the BIM Graduation Design Innovation Competition need to complete the work with the help of the instructors. The competition process involves theoretical knowledge and practical operation. Therefore, the instructors are important for BIM learning performance.

"Teamwork" refers to a method of cooperating with a group of people to achieve a goal through interdependent actions and mutual accountability. Effective teamwork requires individual self-efficacy, proper communication among members, and acceptance of suggestions and instructions from leaders and others (Kossaify et al., 2017). Studies have found that team size and communication frequency between team members have a significant effect on team work performance (Akgün et al., 2005). The National BIM Graduation Design Innovation Competition is a blend of project construction cost, engineering management, building material supply, and so on, which needs students from different majors cooperating with each other and apply different software to complete the solution design, digital construction design, 3D digital model reconstruction, writing reports, and so on. Thus, team members need to communicate effectively and work in close collaboration.

Social environment incentive refers to the incentive effect generated by the social system, production relations, and social relations between people in a specific period and a specific region (Chen Suyue, 2020). External objective factors such as the rigid requirements of schools for students to participate in discipline competitions and the influence of the group they are in, as well as motivational factors such as school rewards and class awards and excellence evaluation, may have a great influence on student' learning performance (Shi et al., 2019). For example, as time goes on, individuals will be influenced by the team, and the goals of individuals and groups will gradually converge (Ladley et al., 2015). At the same time, performance-based rewards have a positive effect on individual performance (Meslec et al., 2020), and rewards can be used as an effective management tool to improve people's work performance (Shaheen et al., 2020).

School is the main platform for education and teaching activities, and the impact of school education and teaching on the students' learning performance is particularly important (Ren Rongchun, 2011). School atmosphere is a comprehensive reflection of the school environment variable, which not only refers to the physical environment in school (such as construction, lighting, and so on), but also to the individual measurement of school experience, including the organizational level, the teaching level, interpersonal relationship, culture, values, and other aspects (Cohen et al., 2009; Nie Qian et al., 2018). The school atmosphere exerts a subtle influence on students, which is directly related to the students' academic achievements and future development (Chunhua, 2014). A good learning atmosphere not only includes reasonable curriculum planning, positive teaching atmosphere, perfect hardware facilities, but also includes the emphasis given by the school to students' extracurricular activities and taking an active part in competitions.

Research shows that college students have a higher understanding of the industry and better academic performance, and they have a higher probability of finding an ideal job (Huang et al., 2017). Therefore, if students have a clearer understanding of their ideal employment after graduation, they will be encouraged to spend more time working towards it. BIM implementation strategies are increasingly driving the use of this technology in the AEC industry worldwide, and Chinese Government agencies have deployed a number of strategies aimed at promoting the development and adoption of BIM (Zhou et al., 2019b). Since 2011, The Ministry of Housing and Urban-Rural Development of China has updated the BIM technology promotion policy annually. In December 2016, the Unified Standard for the Application of Building Information Model approved by the Ministry of Housing and Urban-Rural Development was the first national standard for the application of BIM in China, which had a critical impact on the development of BIM technology in China. From 2017, a series of policies and standard documents issued by the State Council and the Ministry of Transport has also promoted the sustainable development of BIM technology. These policy documents targeted specific projects on the adoption of BIM, and provide technical guidance for the application of BIM across all national engineering endeavors. In 2018, policies and guidelines on BIM implementation were again updated. The construction industry has thus transitioned from traditional construction practices to BIM-based practice. The more students know regarding the application of the BIM within industry, the more they can be expected to be motivated to master BIM technology and in so doing, cultivate the furtherance of BIM uptake across the AEC industry.

5.3 Materials and Methods

5.3.1 Questionnaire Design

Based on the previous research and the tracking investigation of the sixth NBGDIC, the questionnaire design of this study mainly includes two parts: Part I: basic demographic information of the participants, including gender, age, team name, school, major, grade, years of using BIM-related software, the degree of correlation between this competition and their graduation design, and so on; Part II: BIM learning performance driving factor scale. A five-point Likert scale was used to measure the degree of conformity between each observation variable and the contestant [39–43]. Points 1–5 indicated "completely inconsistent", "basically inconsistent", "general", "relatively consistent" and "completely consistent", respectively. The measurement scale of the questionnaire is mainly derived from relevant literature and combined with interviews with competitors. Seven predictors, including individual ability, achievement demand, instructor quality, teamwork, social environment incentive, school (college) atmosphere, and understanding of BIM industry application, were constructed, and 44 corresponding measurement items were designed.

5.3.2 Date Collection

The purpose of the sixth NBGDIC is to promote the BIM application skills for AEC related students. The sixth NBGDIC provides a new way for the application and exploration of BIM technology in colleges and universities, provides ideas for opening BIM-related courses and subject research, and improves the employment rate and employment quality of students. The participating teams of the sixth NBGDIC will upload their work through the Internet according to the registration items. The uploading time is from March 2020 to April 2020. The competition contains seven modules: (A) BIM modeling application, (B) BIM cost management, (C) BIM-based digital construction organization design, (D) BIM-based 5D construction management, (E) structural design based on BIM+ assembly, (F) innovation module, including intelligent construction and management innovation, (G) architectural decoration virtual design. This competition sets up four awards: first prize, second prize, third prize and excellence award. This research questionnaires, as the feedback of the participating students' experience, were issued to the competition system by the sponsor Glodon Company Limited on 8 May 2020. After a month of follow-up investigation, 2790 questionnaires were finally collected, and questionnaires with missing data were eliminated. Finally, 2777 valid questionnaires were collected with an effective recovery rate of 99.53%. The results of the competitors were published on the official website (http://gxbs.glodonedu.com/, May 30, 2020) in late June 2020. In this study, the maximum value of the results for each team was taken as their BIM learning performance, which was assigned as "no winning = 0", "excellence award = 1", "third prize = 2", "second prize = 3", and "first prize = 4".

5.3.3 Data Analysis Method

The SPSS24.0 software was used to analyze the survey data. The Cronbach's alpha coefficient method was used to test the internal consistency of questionnaires data, principal component factor analysis was used to test the structural validity of questionnaires data, and the t-test or ANOVA was used to check the inter-group differences of data. Exploratory factor analysis was conducted on the 44 driving factors of BIM learning performance by maximum variance method. Finally, the BIM competition results are taken as BIM learning performance and the influence of the aforementioned factors based on BIM learning performance is explored by ordered logistics regression (Kuo et al., 2019).

5.4 Results

5.4.1 Data Description

In this study, 2777 valid questionnaires were collected, there were 1582 male contestants and 1195 female contestants. The age group ranged from 19 to 24 years old (n = 2713, accounting for 96.7%). Juniors (n = 1007, 36.3%) and seniors (n = 1340, 48.3%) accounted for more than 70% of the total competitors. The participants came from 330 colleges and universities, among them, 134 were research-oriented colleges/universities, accounting for 4.8%, 2041 were application-oriented colleges/universities, accounting for 73.5%, and 588 were applied research-oriented colleges/universities, accounting for 21.2% (the distribution of the sample colleges/universities are shown in Fig. 5.1). According to different majors of competitors accounting for more than 1%, the competitors were divided into seven categories, including construction management (n = 973, accounting for 35.0%), engineering cost (n = 851, accounting for 30.6%), civil engineering (n = 529, accounting for 19.0%), water supply and drainage science and engineering (n = 71, accounting for 2.6%), building environment and energy application engineering (n = 63, 2.3%), building electrical and intelligent engineering (n = 51, 1.8%), and others (n = 239, 8.6%); the majority of participants have learned and used BIM-related software for no more than 5 years, and those who have used BIM for 2 years are the most (n = 937, accounting for 33.7%). More than 50% of the participants thought that the sixth NBGDIC was related to their graduation designs (n = 1414, accounting for 50.9%); the details of the participants are shown in Table 5.1.

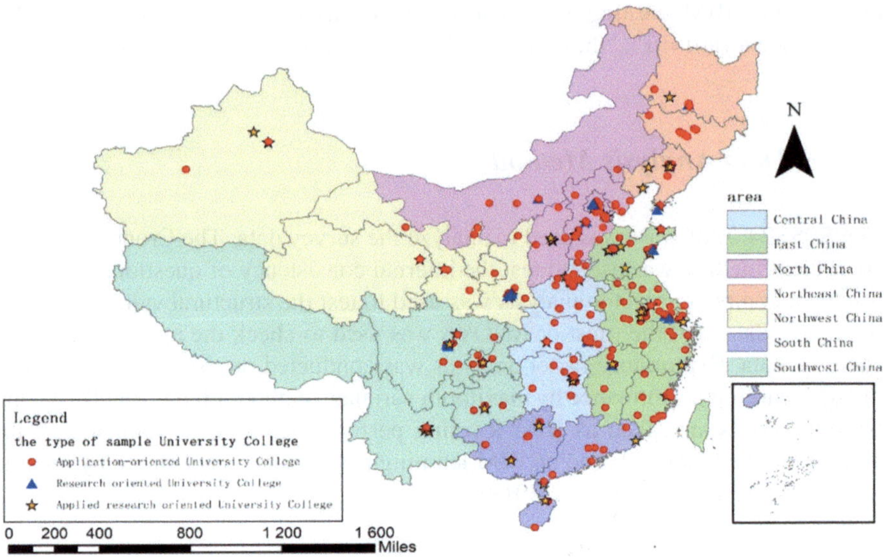

Fig. 5.1 The distribution of the sample colleges and universities

5.4.2 Reliability and Validity Test

Generally, the Cronbach's alpha is greater than 0.8, which indicates an excellent internal consistency of data, which fulfills the requirements for further analysis. The results of the data analysis in this study show that the Cronbach's alpha coefficient is 0.962 and the number of items is 44, indicating that the data have good reliability. Meanwhile, the KMO value of the study data is 0.965, which is greater than the critical value of 0.5, and the significance probability of the Bartlett sphericity test is 0, indicating that the data are suitable for factor analysis.

5.4.3 Difference Analysis Between Groups

Results show that the maximum average score of the contestants is 3.07 ± 1.291, among which 39.4% are above the average score, and 60.6% are below the average score, in which males are (3.1 ± 1.3) and females are (3.0 ± 1.3). The applied research colleges/universities (3.2 ± 1.3) received the highest score, students majoring in building electrical and intelligent engineering received the highest score of (3.3 ± 1.2), and freshmen received the lowest score of (2.5 ± 1.4). Participants who have studied and used BIM-related software for more than 3 years have the highest score (3.3 ± 1.3), and no significant difference exists among the relevant degree of this competition and their graduation design. The difference analysis results of the BIM learning performance between groups are shown in Table 5.2.

Table 5.1 Demographic information of the participants (N = 2777)

Characteristics		Samples	
		Frequency	Proportion (%)
Gender	Male	1582	57.0
	Female	1195	43.0
Age	Under 19	14	0.5
	19–21	1155	41.6
	22–24	1558	56.1
	25 and above	50	1.8
School type	Research	134	4.8
	Application	2041	73.5
	Applied research	588	21.2
Major types	Engineering management	973	35.0
	Engineering cost	851	30.6
	Civil engineering	529	19.0
	Water supply and drainage science and engineering	71	2.6
	Building environment and energy application engineering	63	2.3
	Building electrical and intelligent	51	1.8
	other	239	8.6
Grade	Freshman	41	1.5
	Sophomore	389	14
	Junior	1007	36.3
	Senior	1340	48.3
Time of learning and using BIM software	0–0.5 years (excluding 0.5 years)	734	26.4
	0.5–1 years (excluding 1 years)	937	33.7
	1–2 years (excluding 2 years)	763	27.5
	2–3 years (excluding 3 years)	273	9.8
	3 years or above	70	2.5
The related degree between this competition and their graduation design	Totally irrelevant	46	1.7
	Basically irrelevant	151	5.4
	General	534	19.2
	More relevant	1414	50.9
	Complete related	632	22.8

5.4.4 Exploratory Factor Analysis

In this study, the dimension reduction was conducted on 44 BIM learning driving factor items. In the exploratory factor analysis process, the method of maximum variance was used for factor rotation, which made each factor load change to the

Table 5.2 Analysis of the differences between groups of different categories ($\overline{x} \pm s$)

Characteristics	Category	Achievement maximum
Gender	Male	3.1 ± 1.3
	Female	3.0 ± 1.3
	t	2.682
	P	0.007
Age	Under 19	1.5 ± 0.4
	19–21	1.3 ± 0.0
	22–24	1.3 ± 0.0
	25 and above	1.2 ± 0.2
	F	0.555
	P	0.645
School type	Research	3.1 ± 1.3
	Application	3.0 ± 1.4
	Applied research	3.2 ± 1.3
	F	2.637
	P	0.072
Major types	Engineering management	3.0 ± 1.3
	Engineering cost	3.1 ± 1.3
	Civil engineering	3.2 ± 1.3
	Water supply and drainage science and engineering	3.1 ± 1.2
	Building environment and energy application engineering	3.2 ± 1.5
	Building electrical and intelligent	3.3 ± 1.2
	other	2.8 ± 1.3
	F	2.793
	P	0.01
Grade	Freshman year	2.5 ± 1.4
	Sophomore year	3.1 ± 1.3
	Junior year	3.1 ± 1.3
	Senior year	3.0 ± 1.3
	F	4.284
	P	0.005

(continued)

Table 5.2 (continued)

Characteristics	Category	Achievement maximum
Time of learning and using BIM software	0–0.5 years (excluding 0.5 years)	3.0 ± 1.3
	0.5–1 years (excluding 1 years)	3.1 ± 1.3
	1–2 years (excluding 2 years)	3.1 ± 1.3
	2–3 years (excluding 3 years)	3.2 ± 1.3
	3 years or above	3.3 ± 1.3
	F	2.869
	P	0.022
The related degree between this competition and their graduation design	Totally irrelevant	3.1 ± 1.4
	Basically irrelevant	2.9 ± 1.2
	General	3.0 ± 1.3
	More relevant	3.1 ± 1.3
	Complete related	3.1 ± 1.3
	F	1.587
	P	0.175

two extremes of larger or smaller, and the items with the maximum rotation load of less than 0.4 were deleted. Results show that there were seven common factors with eigenvalues greater than 1, and the total variance explanatory variance was 73.022%. The common factors of the seven dimensions are 1 = the comprehensive ability of instructor, 2 = school atmosphere, 3 = team collaboration, 4 = participants' personal ability, 5 = understanding of BIM industry application, 6 = social environment incentive, and 7 = achievement demand. The specific results of each factor load are shown in Table 5.3.

5.4.5 *Ordered Logistic Regression*

In this study, ordered logistic regression was used to explore the influence of the above factors on BIM learning performance based on the sixth NBGDIC. The results show that among the seven common factors, four common factors significantly affected the BIM learning performance, which were the comprehensive ability of instructor ($p = 0.002 < 0.1$), team collaboration ($p = 0 < 0.1$), participants' personal ability ($p = 0.001 < 0.1$), and achievement needs ($p = 0.066 < 0.01$); Moreover, the gender, type of school, type of major and time of learning and using BIM-related software all have a significant effect on the BIM learning performance; however, the college/university atmosphere, participants' understanding of BIM industry application, social environment incentive, and the correlation between this competition and their graduation design have no significant effect. The detailed results of the ordered logistic regression are shown in Table 5.4.

Table 5.3 Exploratory factor analysis (rotated component Matrix A)

Items	Component						
	1	2	3	4	5	6	7
My instructor always gives me a lot of encouragement	0.856						
My instructor always gives me useful advice	0.855						
My instructor always affirms my progress	0.853						
My instructor will give me in-depth guidance on my entries	0.812						
I get on well with my instructor	0.809						
My instructor attaches great importance to this competition	0.792						
My instructor is very professional	0.788						
My instructor is able to assign tasks according to team members' characteristics	0.759						
Even if my operation is not proficient, my instructor will not blame me	0.748						
My instructor is very familiar with the rules and procedures of this competition	0.741						
My instructor is very influential in the BIM industry	0.599						
If the competition needs, our school can arrange corresponding personnel to coordinate		0.796					
Our school conducted targeted lectures and training for participants		0.768					
Our school values the participation of students in the competition		0.756					
The BIM teaching methods in our school are diverse		0.753					
Our school provides special equipment room and hardware supporting for the participants		0.749					
Our school continuously promotes BIM development to students		0.747					
School teachers have corresponding projects to support BIM technology practice		0.738					
Our school offers a series of BIM related courses		0.728					
School teachers are willing to recommend the winning entries to enterprises		0.726					
School teachers are willing to recommend award-winning students to enterprises		0.723					
Our school/college are willing to provide financial support for participating students		0.687					
Our team has a reasonable division of labor and sincere cooperation			0.805				

(continued)

Table 5.3 (continued)

Items	Component						
	1	2	3	4	5	6	7
Our team has given me a lot of confidence			0.794				
Our team had a smooth communication channel			0.788				
Our team members are willing to share their own experience and technology to each other			0.78				
Our team members have reasonable professional backgrounds			0.74				
I have a strong ability to learn				0.752			
I often learn BIM knowledge by myself on the internet				0.729			
I have solid professional knowledge and theory				0.723			
I am familiar with the rules and procedures of this competition				0.685			
I often participate in BIM related competitions				0.619			
I am familiar with the application of BIM in architectural design					0.798		
I am familiar with the application of BIM in architectural planning					0.772		
I am familiar with the application of BIM in architectural construction					0.77		
I am familiar with the application of BIM in architectural maintenance					0.768		
The purpose of participating in the this competition is to get learning points and other incentives						0.841	
The purpose of participating in this competition is to obtain material rewards						0.836	
It is the instructor's requirement to participate in this competition						0.667	
Influenced by the team, I decided to participate in this competition						0.655	
The purpose of participating in this competition is to improve my personal ability							0.761
The purpose of participating in this competition is to enrich my practical experience							0.749
It is very important to participate in this competition							0.685
It is my interest to participate in this competition							0.62

Note: 1–7 represent the comprehensive ability of instructor, school atmosphere, team collaboration, participants' personal ability, understanding of BIM industry application, social environment incentive, and achievement demand, respectively

Table 5.4 Ordered logistic regression results

		Estimate	The standard error	Wald	Degrees of freedom	Significant	95% a confidence interval Minimum	Maximum
The threshold value	[The maximum = 1]	−1.277	0.215	35.198	1	0	−1.698	−0.855
	[The maximum = 2]	−0.538	0.214	6.353	1	0.012	−0.957	−0.12
	[The maximum = 3]	0.781	0.214	13.332	1	0	0.362	1.2
	[The maximum = 4]	2.106	0.218	93.736	1	0	1.68	2.532
The comprehensive ability of instructor		0.107	0.034	9.641	1	0.002	0.039	0.175
College/university atmosphere		−0.01	0.035	0.078	1	0.78	−0.078	0.059
Team collaboration		0.191	0.034	30.868	1	0	0.123	0.258
Individual ability		0.122	0.036	11.216	1	0.001	0.05	0.193
Understanding BIM application		−0.004	0.035	0.013	1	0.909	−0.072	0.064
Social environment incentive		−0.035	0.034	1.033	1	0.309	−0.102	0.032
Achievement needs		0.063	0.034	3.384	1	0.066	−0.004	0.131
Time of learning BIM related software		0.071	0.036	3.874	1	0.049	0	0.141
The related degree between this competition and their graduation design		−0.008	0.041	0.04	1	0.841	−0.089	0.072
[Gender: male = 1]		0.159	0.073	4.761	1	0.029	0.016	0.301
[Gender: female = 2]		0a			0			
[Type of school: application]		−0.273	0.086	10.1	1	0.001	−0.441	−0.105
[Type of school: research]		−0.38	0.172	4.872	1	0.027	−0.718	−0.043
[Type of school: applied research]		0a			0			
[Major type: engineering management]		0.254	0.132	3.714	1	0.054	−0.004	0.513

	Estimate	The standard error	Wald	Degrees of freedom	Significant	95% a confidence interval	
						Minimum	Maximum
[Major type: engineering cost]	0.359	0.137	6.908	1	0.009	0.091	0.627
[Major type: civil engineering]	0.501	0.141	12.54	1	0	0.224	0.778
[Major type: water supply and drainage science and engineering]	0.472	0.245	3.719	1	0.054	−0.008	0.953
[Major type: building environment and energy application engineering]	0.569	0.255	4.962	1	0.026	0.068	1.07
[Major type: building electrical and intelligent]	0.717	0.279	6.608	1	0.01	0.17	1.264
[Major type: other]	0a			0			

Correlation function: fractional logarithm. A set this parameter to zero because it is redundant

5.5 Discussion

5.5.1 Effects of Demographic Variables

The time of the participants' learning and use of BIM-related software has a significantly positive effect on their BIM learning performance ($\beta = 0.071$, $p = 0.049$). BIM provides a collaboration platform to support collaboration between different business fields (Al-Ashmori et al., 2020), in which competitors need to spend energy and time to learn the operation of BIM software. However, accumulating scientific knowledge can improve innovation performance. Therefore, continuous accumulation of BIM-related software learning and operation is the guarantee to improve BIM learning performance. The BIM learning performance of male students was significantly higher than that of female students ($\beta = 0.159$, $p = 0.029$). BIM learning involves professional theories and software operation practices related to construction. Existing studies show that male students are more in demand in the construction industry, and male students are more suitable to engage in construction-related jobs from the perspective of construction enterprises (Vieira et al., 2017). The BIM learning performance of students from different types of schools is significantly different between groups. The BIM learning performance of students from research-oriented colleges/universities (3.1 ± 1.3) and application colleges/universities (3.0 ± 1.4) is lower than that of students from applied research colleges/universities (3.2 ± 1.3). According to the results of the ordered logistic regression, taking the BIM learning performance of students in applied research universities as the reference, the BIM learning performance coefficients of students in both research and application universities are significantly negative. Therefore, the BIM learning performance of students in applied research colleges/universities is relatively better. BIM learning requires not only a professional theoretical basis, but also practical ability, which is consistent with the educational orientation of "practical application + theoretical research" in applied research universities (Tang, 2020). There are also significant differences in the BIM learning performance of students across different majors. The scores, from high to low, are building electrical and intelligence (3.3 ± 1.2), building environment and energy application engineering (3.2 ± 1.5), civil engineering (3.2 ± 1.3), engineering cost (3.1 ± 1.3), water supply and drainage science and engineering (3.1 ± 1.2), engineering management (3.0 ± 1.3), and others (2.8 ± 1.3). The results of the ordered logistic regression show that the majors of higher education students significantly affect the BIM learning performance. With the continuous promotion and practical application of BIM technology, BIM technology has been widely known by practitioners in the construction industry, and BIM technology education has also been implanted into the talent training programs of AEC-related majors (Li Ming, 2021). In addition, the diversification and complexity of building functions also require the cooperation of multiple professionals to complete the implementation of BIM technology (Zheng, 2020). Therefore, from the analysis of BIM learning performance, students majoring in new technology and construction have better BIM learning performance, but the comprehensive application of BIM technology should be completed in collaboration with multiple majors to realize the effective collaborative management of BIM.

5.5.2 Effects of Common Factors

The above results show that in the sixth NBGDIC, four common factors have significantly positive effects on the BIM learning performance, and the effect degree from large to small is as follows: team collaboration, individual ability, the comprehensive ability of instructor and achievement demand. First, team collaboration is significantly and positively correlated with BIM learning performance ($\beta = 0.191$, $p = 0.000$), indicating that the degree of cooperation between team members will significantly improve BIM learning performance. David W Johnson (1993), an American education scholar, found that "cooperative learning is the most important and most successful in more than ten years" (Johnson, 1993). Cooperative learning has been widely used in teaching reform in many countries and regions in the world. Cooperative learning can improve students' learning performance to the greatest extent (Chen, 2015). As the frequency and quality of team members' communication improves, so does team work performance (Cha et al., 2014). The NBGDIC is a multi-disciplinary cross-integration competition, which requires team members to have a high degree of cohesion and cooperation. Each member should perform a good job in their own professional field and everyone should communicate effectively and work together to complete the competition. Therefore, the degree of team collaboration becomes the most significant factor affecting BIM learning performance. The individual ability of participants significantly affects BIM learning performance ($\beta = 0.122$, $p = 0.001$), which becomes the second most influential factor after team collaboration. Studies have shown that students' self-learning ability and professional foundation significantly affects professional learning performance (Zhou et al., 2019a). Instructor's comprehensive ability also have significant positive effect on BIM learning performance ($\beta = 0.107$, $p = 0.002$). According to the measurement indicators of instructors' comprehensive ability, BIM learning performance can be significantly improved by encouragement, affirmation, effective suggestions and in-depth guidance from instructors. Existing studies have also shown that the harmonious relationship between instructors and students may play an important role in promoting the students' development and affects the students' systematic mastery of certain knowledge (Lopez-Agudo et al., 2020). The results of this study show that participants' achievement requirement have a significant positive effect on their BIM learning performance ($\beta = 0.063$, $p = 0.066$). Existing studies show that the learners' perception of the importance of learning tasks and their interest in learning tasks may significantly affect their learning motivation, attitude, and belief, thus affecting their learning performance (Peiró et al., 2020), which is consistent with our research results.

5.6 Conclusions and Recommendations

This is the first study to take the participants of the Sixth National BIM Graduation Design Innovation Competition of Colleges and Universities as the object of investigation, and to undertake a research design, data collection, and data analysis on the influencing factors of BIM learning performance within such a context. Over the month following the event a questionnaire survey was sent out to participants and a total of 2777 valid responses were returned for analysis. Seven common factors affecting BIM learning performance were extracted using exploratory factor analysis, while an ordered logistic regression was used to ascertain the influence of the derived driving factors on BIM learning outcomes. Findings provide theoretical support for an in-depth understanding of BIM learning drivers, while also offering insights into ways and means by which BIM education in Chinese colleges and universities can be further enhanced. These are:

1. Strengthen the fundamental professional knowledge and sustainable learning ability of college students majoring in AEC related disciplines;
2. Provide real-time attention to the developmental trends of BIM in industry and translate these into the BIM learning environment, which in turn will improve students' incentives, willingness and interest in mastering BIM;
3. Encourage students of multiple majors in AEC related disciplines to cooperate and learn BIM, and create reasonable divisions of labor while sharing skills according to professional backgrounds;
4. Improve the comprehensive ability of BIM instructors in all aspects of BIM, including in competencies such as team liaison, the technical features of BIM, and BIM as a management tool, combined with a systematic understanding of the developmental trends of BIM in the AEC industry;
5. Undergraduate AEC related majors should implant BIM theory and application into current 'talent training' programs across the full duration of the four-year learning process in colleges and universities.

While significant research advancements have been offered in this paper, there are limitations. These may be taken up in subsequent research work:

1. This is the first study on BIM learning performance undertaken with regard to the annual NBGDIC competitions. However, the impact students' various majors have on BIM learning outcomes, while anticipated to exert some influence, was not investigated.
2. In conducting the survey, it was discovered that extant research has paid scant attention to the application of BIM technology in colleges and universities.

The anticipated next related research step will be to examine how to fully combine BIM education with university education, promote and publicize BIM education in colleges and universities, promote BIM as a legitimate, independent professional course in colleges and universities, and conduct in-depth research on the development of cooperative ties regarding BIM between universities and industry.

References

AboWardah, E. S. (2020). Bridging the gap between research and schematic design phases in teaching architectural graduation projects. *Frontiers of Architectural Research, 9*(1), 82–105. https://doi.org/10.1016/j.foar.2019.04.005

Adams, R. J. (2005). Reliability as a measurement design effect. *Studies in Educational Evaluation, 31*(2), 162–172. https://doi.org/10.1016/j.stueduc.2005.05.008

Akgün, A. E., Byrne, J., Keskin, H., Lynn, G. S., & Imamoglu, S. Z. (2005). Knowledge networks in new product development projects: A transactive memory perspective. *Information & Management, 42*(8), 1105–1120. https://doi.org/10.1016/j.im.2005.01.001

Al-Ashmori, Y. Y., Othman, I., Rahmawati, Y., Amran, Y. H. M., Sabah, S. H. A., Rafindadi, A. D. u., & Mikić, M. (2020). BIM benefits and its influence on the BIM implementation in Malaysia. *Ain Shams Engineering Journal, 11*(4), 1013–1019. https://doi.org/10.1016/j.asej.2020.02.002

Alizadehsalehi, S., Hadavi, A., & Huang, J. C. (2020). From BIM to extended reality in AEC industry. *Automation in Construction, 116*(C), 103254.

Alizadehsalehi, S., Hadavi, A., & Huang, J. C. (2021). Assessment of AEC students' performance using BIM-into-VR. *Applied Sciences-Basel, 11*(7), Article 3225. https://doi.org/10.3390/app11073225

Buelow, M. T., & Cayton, C. (2020). Relationships between the big five personality characteristics and performance on behavioral decision making tasks. *Personality and Individual Differences, 160*. https://doi.org/10.1016/j.paid.2020.109931

Cha, M., Park, J.-G., & Lee, J. (2014). Effects of team member psychological proximity on teamwork performance. *Team Performance Management, 20*(1/2), 81–96. https://doi.org/10.1108/tpm-03-2013-0007

Chen, R. (2015). The effect of cooperative learning on primary school students' academic performance. *Basic Education, 12*, 105–112. https://doi.org/10.3969/j.issn.1005-2232.2015.04.013

Chen Suyue, C. M. (2020). Study on the influencing factors of college students' participation in discipline competitions. *Circum-Bohai Economic Outlook*, (7), 141–143.

Chen, K. Y., Chen, W. W., Cheng, J. C. P., & Wang, Q. (2020). Developing efficient mechanisms for BIM-to-AR/VR data transfer. *Journal of Computing in Civil Engineering, 34*(5), 04020037. https://doi.org/10.1061/(asce)cp.1943-5487.0000914

Cheng, J. C. P., Chen, W. W., Chen, K. Y., & Wang, Q. (2020). Data-driven predictive maintenance planning framework for MEP components based on BIM and IoT using machine learning algorithms. *Automation in Construction, 112*, Article 103087. https://doi.org/10.1016/j.autcon.2020.103087

Chunhua, Z. (2014). The impact of school climate on students' academic performance: based on the analysis of 2012PISA results. *Foreign Primary and Secondary Education, 8*, 11–20.

Cohen, J. C., McCabe, E. M., Michelli, N. M., & Pickeral, T. (2009). School climate: Research, policy, practice, and teacher education. *Teachers College Record, 111*(1), 180–213.

Coupry, C., Noblecourt, S., Richard, P., Baudry, D., & Bigaud, D. (2021). BIM-based digital twin and XR devices to improve maintenance procedures in Smart buildings: A literature review. *Applied Sciences-Basel, 11*(15), article 6810. https://doi.org/10.3390/app11156810

Eika, M., Espnes, G. A., & Hvalvik, S. (2014). Nursing staff's actions during older residents' transition into long-term care facility in a nursing home in rural Norway [Article]. *International Journal of Qualitative Studies on Health and Well-Being, 9*(12), Article 24105. https://doi.org/10.3402/qhw.v9.24105

Escribano, R., Treviño, E., Nussbaum, M., Torres Irribarra, D., & Carrasco, D. (2020). How much does the quality of teaching vary at under-performing schools? Evidence from classroom observations in Chile. *International Journal of Educational Development, 72*. https://doi.org/10.1016/j.ijedudev.2019.102125

Fichter, L., von Bernstorff, C., & Rohrmann, S. (2020). The necessity of targeted aptitude diagnostics for company founders: relationships between personality traits, perceived stress, satisfaction, and performance. *Heliyon, 6*(9), e04987. https://doi.org/10.1016/j.heliyon.2020.e04987

Fleming, J., Gibson, R., Anderson, M., & Martin, A. J. (2020). Quality teaching frameworks and arts education: Seeking a compatible approach. *Journal of Curriculum Studies*, 1–17. https://doi.org/10.1080/00220272.2020.1720299

Hotulainen, R., Vinni-Laakso, J., & Kupiainen, S. (2020). Development of learning to learn competence across secondary education and its association with attainment in Finnish/Swedish high-stake exit exam. *Thinking Skills and Creativity*. https://doi.org/10.1016/j.tsc.2020.100738

Huang, X.-W., Ren, N., & Shen, X.-M. (2017). An empirical study on influencing factors of college students' employment. *Modern Trade and Industry, 38*(11), 77–78.

Ingvarson, L. (2019). Teaching standards and the promotion of quality teaching. *European Journal of Education*. https://doi.org/10.1111/ejed.12353

Jie, J. (2020). Research on the second classroom teaching mode of BIM driven by discipline competition. *Modernization of Education, 52*, 70–73. https://doi.org/10.16541/j.cnki.2095-8420.2020.52.020

Johnson, D. W. (1993). Cooperation in the classroom.

Kossaify, A., Hleihel, W., & Lahoud, J. C. (2017). Team-based efforts to improve quality of care, the fundamental role of ethics, and the responsibility of health managers: monitoring and management strategies to enhance teamwork. *Public Health, 153*, 91–98. https://doi.org/10.1016/j.puhe.2017.08.007

Kuo, C. I., Wu, C.-H., & Lin, B.-W. (2019). Gaining from scientific knowledge: the role of knowledge accumulation and knowledge combination. *R&D Management, 49*(2), 252–263.

Ladley, D., Wilkinson, I., & Young, L. (2015). The impact of individual versus group rewards on work group performance and cooperation: A computational social science approach. *Journal of Business Research, 68*(11), 2412–2425. https://doi.org/10.1016/j.jbusres.2015.02.020

Li Ming, L. N. (2021). Practice of BIM teaching scheme for application oriented talents training. *The International Journal of Electrical Engineering Education, 60*, 002072092199660.

Liu, H.-Y., Chao, C.-Y., Kain, V. J., & Sung, S.-C. (2019). The relationship of personal competencies, social adaptation, and job adaptation on job satisfaction. *Nurse Education Today, 83*. https://doi.org/10.1016/j.nedt.2019.08.017

Lopez-Agudo, L. A., Marcenaro-Gutierrez, O. D., & Molina-Marfil, J. A. (2020). School tutoring and academic performance: A too close relationship? *Studies in Educational Evaluation, 66*. https://doi.org/10.1016/j.stueduc.2020.100903

Lu, G., C., L., He, Q., & Yan, H. (2018). Evaluation of discipline competition in colleges and universities: Thoughts, methods and exploration. *Higher Education Research in China, 74*(2), 63–68. https://doi.org/10.16298/j.cnki.1004-3667.2018.02.12

Lv, H., Z., L., Zhang, Z., & Shang, L. (2020). Exploration on the deep integration of discipline competition and experimental center teaching reform. *Laboratory Research and Exploration, 39*(1), 170–172.

Ma, X., Chan, A. P. C., Li, Y., Zhang, B., & Xiong, F. (2020). Critical strategies for enhancing BIM implementation in AEC projects: Perspectives from Chinese practitioners. *Journal of Construction Engineering and Management, 146*(2). https://doi.org/10.1061/(asce)co.1943-7862.0001748

Meslec, N., Curseu, P. L., Fodor, O. C., & Kenda, R. (2020). Effects of charismatic leadership and rewards on individual performance. *The Leadership Quarterly*. https://doi.org/10.1016/j.leaqua.2020.101423

Nie Qian, Z. D., Luo, X., Li, Q., & Teng, Z. (2018). The effect of school atmosphere on academic achievement and its theory. *Education Science Forum, 14*, 11–15.

Niebuhr, R. E. A. U., School of Business, & Norris, D. R. (1982). The influence of individual characteristics on performance under varying conditions. *Journal of Social Psychology, 117*(2), 249. https://doi.org/10.1080/00224545.1982.9713435

Peiró, J. M., Bayona, J. A., Caballer, A., & Di Fabio, A. (2020). Importance of work characteristics affects job performance: The mediating role of individual dispositions on the work design-performance relationships. *Personality and Individual Differences, 157*. https://doi.org/10.1016/j.paid.2019.109808

Ren Rongchun, X. T. (2011). The role of schools in the process of equity in educational outcomes – Based on the perspective of the impact of schools on student achievement. *Basic Education, 2*, 30–35.

Shaheen, K., Waheed, A., & H. Hashmi, W. (2020). Extrinsic rewards and creative performance syndrome: The mediating mechanism and interacting effects. *Thinking Skills and Creativity, 38*. https://doi.org/10.1016/j.tsc.2020.100713

Shi, M., Shi, L., & Ge, F. (2019). Analysis of influencing factors of college students' participation in discipline competitions based on questionnaire survey. *Journal of Hubei Second Normal University, 36*(12), 78–82.

Tang, Y. (2020). Exploration on engineering practice and innovation ability cultivation of civil engineering major in Applied Research University. *Literature of Science and Education, 11*, 63–64. https://doi.org/10.16871/j.cnki.kjwhb.2020.04.030

Vieira, A., Carrieri, A. D. P., Monteiro, P. R. R., & Roquete, F. F. (2017). Gender differences and professional identities in health and engineering. *BAR – Brazilian Administration Review, 14*(1). https://doi.org/10.1590/1807-7692bar2017160082

Wang Yana, J. L., & Mao Hanping. (2019). Analysis on the influencing factors of college students' innovation performance in discipline competition. *University Education Management, 13*(5), 104–114.

Wang, L., Huang, M., Zhang, X., Jin, R., & Yang, T. (2020). Review of BIM adoption in the higher education of AEC disciplines. *Journal of Civil Engineering Education, 146*(3).

Yongliang Deng, J. L., Qiuting Wu, Shuangshuang Pei, Xu, N., & Ni, G. (2020). Using network theory to explore BIM application barriers for BIM sustainable development in China. *Sustainability, 12*(8), 1–22.

Zheng, Q. (2020). Development and application of building electrical and intelligent building. *Mass Standardization, 9*, 95–96.

Zhou, W., Min, Q., & Guan, Y. (2019a). An empirical study on the influencing factors of undergraduate teaching effect. *University Education Management, 13*, 104–112.

Zhou, Y., Yang, Y., & Yang, J.-B. (2019b). Barriers to BIM implementation strategies in China. *Engineering Construction and Architectural Management, 26*(3), 554–574. https://doi.org/10.1108/ecam-04-2018-0158

Zhou, P., Heng, L. I., Hong-Yu, L. I., He, Y., & Weerasinghe, R. (2020). Applications of BIM in construction engineering in China: A review. *E3S Web of Conferences, 143*. https://doi.org/10.1051/e3sconf/202014301007

Chapter 6
What Determines BIM Competition Results of Undergraduate Students in the Architecture, Engineering and Construction Industry?

Abstract Building Information Modeling (BIM) has become the development trend in the architecture, engineering, and construction (AEC) industry. BIM discipline competition is an effective way for students of AEC-related disciplines to integrate theory with practice, and it is a key link to cultivate qualified BIM practitioners. This study takes participants of the eighth National College BIM Graduation Design Innovation Competition as the research objects. The Technology Acceptance Model and the Theory of Planned Behavior are combined to build a driving factor model of the competitions. Contestants of the competition were asked to complete online questionnaires, and 451 valid samples were finally obtained. Structural Equation Modeling was used to fit the theoretical model, and it was found that: Behavioral Intention to use BIM is directly and positively affected by Attitude, Perceived Behavioral Control, and Perceived Usefulness, as well as indirectly and positively affected by Perceived Usefulness and Perceived Eased of Usefulness. Competition Results is directly and positively affected by Behavioral Intention to use BIM, Perceived Behavioral Control, and Facilitating Conditions. It is also indirectly and positively affected by Attitude, Perceived Behavior Control, Perceived Usefulness, and Perceived Eased of Usefulness. The results show that the situation of students' participation in BIM competition can be optimized by increasing the publicity and promotion of BIM and related policies, strengthening the construction of supporting teaching facilities, building the integrated curriculum system of BIM technology, and strengthening teacher training. This study provides a theoretical reference for further BIM practice, and it would help improve the corresponding teaching organization and enhance the internal drive of students' BIM learning.

Keywords BIM · Technology acceptance model · Theory of planned behavior · Behavioral intention · Competition results

Y. Ao et al., *Building Information Modeling (BIM) Adoption*, Resilient and Friendly Living Environments, https://doi.org/10.1007/978-981-95-3993-2_6

6.1 Introduction

The construction industry has changed from traditional construction practice to BIM-based practice (Ao et al., 2021). As the second revolution of the construction industry, BIM technology has effectively promoted the sustainable development of the construction industry (Atabay et al., 2020). However, most construction companies still do not have a holistic understanding of BIM (Al-Ashmori et al., 2020), and there are still problems in the implementation of BIM technology. BIM's implementation difficulties include poor government leadership, insufficient guidance, organizational problems, legal problems, high application costs, and insufficient external incentives (Zhou et al., 2019). The demand for BIM technical talents is urgent, and the development of the industry also forces colleges to offer BIM-related courses to cultivate BIM technical talents. In recent years, colleges and universities in China have successively set up BIM courses; however, the teaching effect is not ideal due to such bottlenecks as "the disconnect between teaching and practice", "students' lack of ability to solve problems independently", "students' lack of innovation practice", "lack of experienced teachers", and "poor hardware conditions" (Gao et al., 2021; Xu et al., 2018; Zhang et al., 2018; Zhao et al., 2020). Colleges and universities must review the gap between BIM education and industry practice (Ao et al., 2021). With the government's strong support, some regions and enterprises have conducted various BIM competitions, hoping to stimulate students' ability to connect theory with practice, which is lacking in the current BIM education (SU & ZHANG, 2021; Wang et al., 2021). However, more studies on the influencing factors of BIM competition and the effect of BIM competitions on the improvement of students' comprehensive ability are needed to provide theoretical and empirical evidence to improve BIM competition and education.

Therefore, this study aims to understand the role of discipline competition in promoting students' learning of BIM technology, software application, and mastery of professional knowledge. Taking the eighth National College BIM Graduation Design Innovation Competition as the platform, this study constructs the driving factors model based on the integrative application of the Technology Acceptance Model (TAM) and the Theory of Planned Behavior (TPB). Perceived Eased of Usefulness (PEU), Perceived Usefulness (PU), Attitude (AT), Subjective Norm (SN), Perceived Behavioral Control (PBC), Facilitating Conditions (FC), Behavioral Intention to use BIM (BI), and Competition Results (CR) are involved in the model to explore factors that affect the contestants' competition results and their willingness to continue using BIM technology. This study intends to provide a theoretical basis for further improvement of the BIM discipline competition quality and BIM education quality.

6.2 Literature Review

6.2.1 Application of BIM

BIM is meant for the process of creating and managing information, which enables AEC professionals to effectively plan, design, construct, and manage the whole construction project's life cycle (Autodesk, 2021). With BIM's implementation, project management's complexity and difficulty will be significantly reduced (He et al., 2017), and productivity will be improved (Al-Ashmori et al., 2020). Many countries have been committed to promoting BIM technology. For example, since 2016, South Korea has required that all public buildings with a cost of more than USD 27.6 million must introduce BIM technology (Zhou et al., 2019), and the British government formulated a strategic policy, expecting to achieve the complete collaboration of BIM in all public sector projects by 2016 (Howard et al., 2017). However, the application of BIM has not reached the expected level. There are some problems in the practice process, such as insufficient understanding of BIM, lack of training, and ambiguous roles and responsibilities (Howard et al., 2017). Howard et al. (2017) found that current policies and research are mainly concentrated at the industry, enterprise, and project level, while the needs at the individual participant level have not gotten enough attention. Al-Ashmori et al. (2020) found that most construction companies in Malaysia lack an understanding of BIM technology.

Compared with developed countries, the development of BIM technology in China has been initiated relatively late, but the policy support of the Chinese government for BIM technology has been continuous. In the "14th five-year plan for the development of housing and urban, rural construction science and technology", the importance of BIM technology is emphasized again. The document mentioned to "study the theory, method, and support system of the integrated application of BIM technology and the new generation of information technology to support the digital transformation and development of the construction industry" (Ministry of Housing and Urban-Rural Development, 2022). Heng and Hong-yu (2020) divided the research of BIM technology in China into two stages. The first stage is from 2002 to 2014 when scholars understand BIM in the primary stage. The second stage starts after 2015, during which BIM-related research is in the ascendant and has made some progress. Zhou et al. (2019) divided the strategies prepared in China into five categories: BIM standards and guides, BIM software development, BIM training for talents, BIM-related academic research, and BIM best practices. With the vigorous promotion of the Chinese government, many construction projects have adopted BIM technology. However, there are still obstacles in construction projects, such as insufficient government guidance, insufficient external incentives, the unclear willingness of project participants to cooperate, a lack of professionals, and others (Heng & Hong-yu, 2020; Su & Zhang, 2021; Zhou et al., 2019). At present, the adoption of BIM technology still lags behind expectations.

6.2.2 Teaching of BIM

The improvement of BIM education will positively impact the BIM awareness (Atabay et al., 2020). BIM modeling in architectural education is conducive to developing students' project concepts and design ability (Çakır & Uzun, 2020). Scholars have studied how to integrate BIM into the existing curriculum system and also studied problems in BIM education. Elgewely et al. (2021) proposed to integrate BIM into architectural engineering education. They confirmed through empirical research that using Virtual Reality (V.R.) to realize students' practical teaching can significantly improve their learning experience and performance. Clevenger et al. (2007) studied the way to integrate BIM technology into architectural engineering-related courses. Casasayas et al. (2021) divided reasons that hinder BIM education in Australian universities into four aspects: change management adjustment, curriculum and content restrictions, problems of educators, and disconnection from the industry. Through the research, they pointed out that researchers should provide the link between graduates' preparation for BIM and their employability in the Australian market. They also argued that relevant research should highlight the industry's needs. In addition, Puolitaival and Forsythe (2016) proposed that the preparation and optimal use of practical models in high-quality education is a major challenge.

The training of BIM technical talents in China is still in infancy (Sun & Cao, 2021), and there are many problems, such as "lack of equipment and facilities", "unsystematic curriculum setting", "the teachers' insufficient professionalism", and "lack of systematic curriculum application" (Jia, 2021). To improve the teaching effect of BIM technology, some colleges integrate BIM technology into the graduation design, hoping to strengthen the connection between BIM education and industry (Li, 2021).

6.2.3 Discipline Competition

In order to make up for the problems of disconnection from actual projects and insufficient practical experience of students in BIM education, various competitions have been held in collaboration with industries, enterprises, and colleges. Due to a large number of practical engineering projects, the participation of enterprises in competitions can significantly improve the connection between students and practical engineering (Jiang, 2020). With the development of various competitions, scholars have conducted research on discipline competitions, mainly involving the pattern and problems of competitions, the combination of competition and school curriculum, and the impact on students' personal ability (Fu, 2018; Lu et al., 2018; Plakhotnik et al., 2020; Shi et al., 2019; Wang et al., 2018; Yang & Ha, 2018). Plakhotnik et al. (2020) confirmed that the experience of participating in competitions significantly impacts the career decision-making and self-efficacy of participants. With the joint promotion of universities and enterprises, discipline competitions in China are booming. However, there are also some problems, such as "repeated content", "limited project types", and "a limited number of trainees" (Lu et al., 2018). Currently, most of the research on discipline competition focuses on the innovative talent training mode and the operation status analysis. However, the internal influencing factors of students' participation in discipline competitions have not been thoroughly studied. However, students are the main body of discipline competitions, so analyzing the internal influence in discipline competitions is necessary (Shi et al., 2019).

The existing studies are mainly carried out from the macro perspective, while more attention is needed at the personal level. The cultivation mode of BIM technical talents in colleges are still being explored, and there is no widely recognized system mode so far. Although the role of BIM discipline competitions in promoting BIM talent training has been demonstrated, there is less understanding of the competition's effectiveness from the contestants' perspective. Therefore, this study takes contestants as the object to study the influencing mechanism of the contestants' competition results, expecting to provide a theoretical basis for improving BIM competition and BIM education.

6.3 Methodology

6.3.1 Research Theoretical Model

Davis et al. (1989) proposed the TAM in 1989, which points out that the behavior of people facing new things will be affected by PEU, PU, AT, and BI (Sun & Cao, 2021). Many scholars have used the TAM model to study the construction industry's acceptance of BIM technology. Acquah et al. (2018) used TAM to test the acceptance of BIM in Ghana's construction industry, while Lee et al. (2015) developed a BIM acceptance model based on the theory related to technology acceptance behavior. In addition, Zhao et al. (2022) studied the specific mechanism of Chinese AEC professionals' acceptance of BIM technology in combination with the TAM and Technology Organization Environment model.

Scholars also attempt to integrate various theoretical models to better explain people's acceptance behavior in the face of new technologies. Venkatesh et al. (2003) proposed the Unified Theory of Acceptance and Use of Technology (UTAUT) in 2003, which integrates eight main models: the Theory of Reasoned Action, the Technology Acceptance Model, the Motivation Model, the Theory of Planned Behavior, the combined TAM-TPB model, the Model of Personal Computer Utilization, the Innovation Diffusion Theory, and Social Cognition Theory.

In addition, TPB, as one of the most widely used theories in social and behavioral sciences, has been widely used in health science, environmental science, educational research, and other fields (Bosnjak et al., 2020). TPB mainly proposes that behavior intention is formed by three factors: attitude, subjective norms, and perceived behavior control. Some scholars also use TPB to explore the final use behavior of BIM technology. For example, Z. Wu et al. (2021) confirmed the applicability of the TPB with the study of critical factors of professionals' BIM technology adoption behavior.

Generally, while participants have a firm behavior intention on BIM technology, they will have stronger motivation in the process of the competition to achieve better competition results. Howard et al. (2017) verified the significant impact of behavioral intention on behavior in empirical research. When people are facing new technologies, their behavior depends on their attitude towards new technologies (Su & Zhang, 2021). Botero et al. (2018) confirmed that attitude is a powerful factor affecting behavioral intention. Zhou and Li (2022) verified that the acceptability of secondary vocational students to the mixed teaching model would be affected by the people around them, while Venkatesh and Zhang (2014) also verified that the influence of the surrounding environment and people can significantly impact individual behavior through comparing the social environment of China and the United States. In UTAUT, it is believed that the promotion conditions will also affect people's final use behavior (Venkatesh & Zhang, 2014). In addition, Ao et al. (2021) pointed out that the results of students participating in competitions will be affected by the guidance of instructors. Therefore, the factor of "Facilitating Conditions" has also been considered in this study.

The variables in this study model are defined as follows.

CR: The final performance of a competitor in a BIM competition.
FC: How much support the individual feels from the organization for learning BIM technology and participating in the competition in terms of relevant technologies and equipment.
BI: Individual's subjective judgment on the probability of continuing to learn BIM technology and engaging in relevant industries in the future.
AT: Individual's positive or negative feelings about learning BIM technology and participating in the competition.
SN: The degree to which an individual perceives others around him as to whether he should learn BIM technology and enter the competition.
PU: The extent to which an individual believes learning BIM technology and participating in the competition will help his or her abilities or future career.
PBC: The extent to which individuals anticipate learning BIM technology and participating in the games that they can control or master.
PEU: How easy an individual thinks it is to learn BIM and compete.

Based on the above research conclusions and the team's research basis for the BIM competition, this study puts forward the following assumptions:

Hypothesis 1 (H1): AT of participants has a positive effect on BI;
Hypothesis 2 (H2): SN of participants has a positive effect on BI;
Hypothesis 3 (H3): PBC of participants has a positive effect on BI;
Hypothesis 4 (H4): BI has a positive impact on CR;
Hypothesis 5 (H5): PBC of participants has a positive effect on CR.

When participants feel that learning BIM technology is not difficult and that BIM technology benefits the growth of their comprehensive personal ability, their attitude towards BIM technology may be more positive. Their intention to use BIM technology in the future may also be enhanced. Y Sun and S Cao (2021) verified that when students feel that learning BIM technology is helpful, they will learn BIM technology with a positive attitude, and their subsequent behavior will be affected. Meanwhile, when the difficulty of learning BIM technology is acceptable, students would think that learning and using BIM technology is useful, and their corresponding learning behavior would be promoted. Zhao et al. (2022) also verified that perceived usefulness and perceived eased of usefulness have a significant positive effect on attitude. Therefore, this study proposes the following assumptions:

Hypothesis 6 (H6): PU has a positive effect on AT;
Hypothesis 7 (H7): PEU has a positive effect on AT;
Hypothesis 8 (H8): PEU has a positive effect on PU;
Hypothesis 9 (H9): PU has a positive effect on BI.

When participants can get adequate support, such as equipment and technical guidance, they would solve problems in a timely and effective manner in the competition, which may encourage them to achieve better competition results. Howard et al. (2017) also confirmed that good external promotion conditions significantly positively impact the final use behavior. Therefore, the following assumptions are made:

Hypothesis 10 (H10): FC has a positive effect on CR.

According to the above research assumptions, the systematic theoretical model proposed in this study is shown in Fig. 6.1.

6.3.2 Data Collection

According to the above theoretical model, the questionnaire has been implemented to collect data. The questionnaire is mainly divided into two parts: the first part is the demographic variables of the contestants, including gender, grade, major, and others, and the second part is all the measurement items involved in the model. The five-point Likert scale was used to measure the degree of conformity between each observation variable and the contestant (1-completely disagree; 2-basically disagree; 3-average 4-basically agree; 5-completely agree). The specific items of the questionnaire are shown in Table 6.1.

The eighth National College BIM Graduation Design Innovation Competition, as one of the BIM competitions with high recognition, was held from October 2021 to May 2022. The competition provides a new pattern for the application exploration of BIM education and the curriculum of BIM-related courses. The competition includes nine aspects: civil construction BIM modeling and application, mechanical and electrical BIM modeling and application, the whole process cost management and application, BIM bidding management and application, BIM decoration design

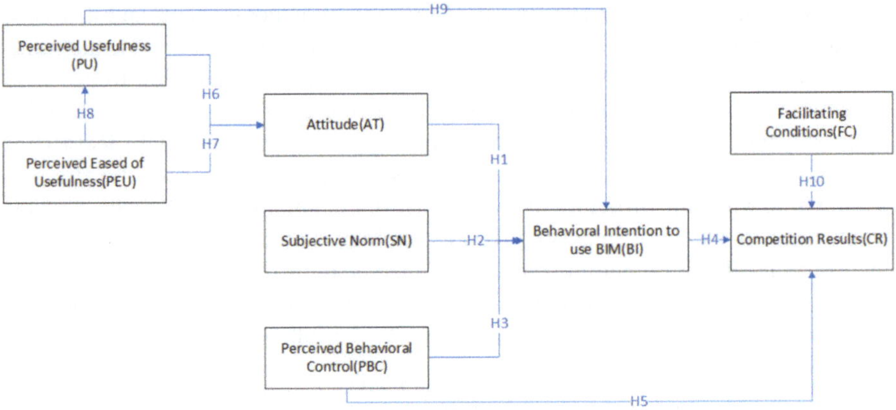

Fig. 6.1 Driving factors model of BIM competition results based on TAM-TPB

creativity and innovation, application of BIM construction project management, prefabricated building BIM design and construction, intelligent building and management innovation and BIM positive design application innovation. There are four prizes: first prize, second prize, third prize, and excellence award. More than 560 colleges and universities have joined the competition. This research is divided into two stages. The first stage was the pre-survey conducted from 28 March 2022 to 3 April 2022. After the pre-survey, the questionnaire is adjusted. The second stage is the formal investigation, conducted from 5 April to 28 May 2022. The questionnaire collection was completed with the assistance of the organizer, Glodon Company Limited, of the competition. The questionnaires were distributed to the participating students through the system during the competition. A total of 828 questionnaires were collected, of which 451 were valid, with an effective rate of 54.47%. Table 6.2 represents the interviewees' statistics.

6.4 Results

6.4.1 Reliability and Validity

Cronbach's α coefficient is used to test the reliability of the questionnaire. It is generally believed that the questionnaire has good reliability when the Cronbach's α coefficient reaches 0.8 (Wu et al., 2021). The overall Cronbach's α coefficient of the questionnaire was 0.959. Additionally, the Cronbach's α coefficient of PU, PEU, SN, FC, AT, PBC, and BI, respectively, are 0.948, 0.849, 0.930, 0.912, 0.915, 0.910, and 0.931, all of which reached above 0.8, indicating good reliability of the questionnaire.

Confirmatory factor analysis was used to test the validity of the questionnaire, and the test results showed that the KMO value was 0.947, which was higher than

Table 6.1 Measuring Scale

Latent variables	Codes	Observed variables	Source
PEU	PEU1	Participating in BIM competition can improve professional knowledge	Davis et al. (1989); Venkatesh et al. (2003); Zhao et al. (2022)
	PEU2	Participating in BIM competition can improve team cooperation ability and communication ability	
	PEU3	Learning BIM technology is helpful for my career development	
PU	PU1	BIM competition is not difficult for me	Davis et al. (1989); Venkatesh et al. (2003); Zhao et al. (2022)
	PU2	The interface design of BIM software is reasonable and easy to operate	
	PU3	BIM software is easy to learn and easy to use	
SN	SN1	People who affect my behavior think I should participate in the BIM competition	Ajzen (1991); Neto et al. (2020); Venkatesh et al. (2003); Wu et al. (2021)
	SN2	People who are important to me think I should participate in the BIM competition	
	SN3	People who affect my behavior think I should learn BIM technology	
PBC	PBC1	Even if there is no help, I am confident to complete the BIM graduation design competition	Ajzen (1991); Neto et al. (2020); Venkatesh et al. (2003); Wu et al. (2021); Zhang et al. (2018)
	PBC2	Even if I have no BIM competition experience before, I am confident to complete the BIM graduation design innovation competition	
	PBC3	As long as I try my best, I can always solve problems encountered in the BIM competition	
FC	FC1	When I encounter difficulties during the competition, instructors/classmates can help me solve difficulties	Ao et al. (2021); Venkatesh et al. (2003); Venkatesh and Zhang (2014)
	FC2	BIM technology is related to my major	
	FC3	My school supports us in participating in BIM competitions	
AT	AT1	Learning BIM technology is interesting	Howard et al. (2017); Sun and Cao (2021)
	AT2	It's fun to participate in the BIM competition	
	AT3	I like to participate in the BIM competition	
BI	BI1	I intend to continue to study/use BIM technology	Fishbein and Ajzen (1975); Plakhotnik et al. (2020)
	BI2	If I am engaged in the construction industry in the future, I will give priority to BIM-related positions	
	BI3	I would pay more attention to BIM-related positions while looking for part-time internships and employment	

Table 6.2 Demographic distribution

Characteristics		Sample	
		Frequency	Proportion (%)
Gender	Male	251	55.7
	Female	200	44.3
Age	Under 19	9	2.0
	19–21	256	56.8
	22–24	179	39.7
	25 and above	7	1.5
Major	Engineering management	125	27.7
	Engineering cost	146	32.4
	Civil engineering	100	22.2
	Water supply and drainage science and engineering	7	1.6
	Building environment and energy application engineering	10	2.2
	Building electrical and intelligent	8	1.8
	Other	55	12.2
Grade	Freshman	13	2.9
	Sophomore	116	25.7
	Junior	197	43.7
	Senior	125	27.7

the critical value of 0.7 (Wang et al., 2018), and the p-value of significance was 0. Meanwhile, all factor loadings were between 0.629 and 0.924, which were at acceptable levels. The AVE was calculated to test the convergent validity of the questionnaire data. If the value was greater than 0.5, the test was passed; if the CR was greater than 0.7, the test was passed (Fornell & Larcker, 1981). The test results are shown in Table 6.3, which shows that the results are good and can be further analyzed in the structural equation model.

6.4.2 *Goodness of Fit*

The fitting index of the structural equation model is shown in Table 6.4. All the indexes meet the recommended optimal value requirements except GFI. Although GFI does not meet the requirements of the optimal recommended value, it is also very close to the required 0.9, and it is also considered to be within the range of acceptability when GFI is greater than 0.8 (Wu et al., 2021), so it could be considered that the model has a good fitting. The model fitting results are shown in Table 6.4.

Table 6.3 Test results of the measurement model

Variable	CR	AVE	Distinctions						
			PU	PEU	SN	PBC	FC	AT	BI
PU	0.949	0.861	0.928						
PEU	0.8533	0.659	0.775	0.812					
SN	0.932	0.822	0.518	0.561	0.906				
PBC	0.912	0.776	0.578	0.572	0.292	0.881			
FC	0.913	0.778	0.708	0.731	0.366	0.799	0.882		
AT	0.919	0.791	0.771	0.724	0.371	0.728	0.767	0.889	
BI	0.931	0.819	0.709	0.719	0.335	0.664	0.796	0.763	0.905

Table 6.4 Structural model fitting index

Index	CMIN/DF	GFI	AGFI	RMSEA	NFI	RFI	IFI	TLI	CFI
Text value	2.878	0.896	0.864	0.065	0.944	0.934	0.963	0.956	0.963
Recommended value	1 ~ 3	>0.9	>0.9	<0.08	>0.9	>0.9	>0.9	>0.9	>0.9
Source	Wu et al. (2021); Zhao et al. (2022)								

6.4.3 Path Analysis

According to the empirical research results, except for hypothesis 2, all other hypotheses are accepted. The fitting results of the model are shown in Fig. 6.2. The results of hypothesis testing are shown in Table 6.5.

6.4.4 Total Effect, Direct Effect, and Indirect Effect

In order to clearly reflect the total effect, direct effect, and indirect effect among variables, the methods of "indirect, direct, and total effect" are used when exporting the results. The total effect, direct effect, and indirect effect among variables are shown in Table 6.6. The relationship between direct and indirect effect variables is shown in Fig. 6.3.

6.5 Discussion

PEU has a significant positive impact on CR (0.122). As can be seen from Table 6.6 and Fig. 6.2, the action path of PEU on CR can be divided into three aspects: PEU → AT → BI → CR, PEU → PU → AT → BI → CR, PEU → PU → BI → CR. PEU had directly significant positive effects on AT (0.541) and PU (0.834). Meanwhile, PEU has an indirectly significant positive effect on AT (0.267) through "PU → AT (0.321)", which strengthens the total effect of PEU on AT (0.808). Similarly, PEU

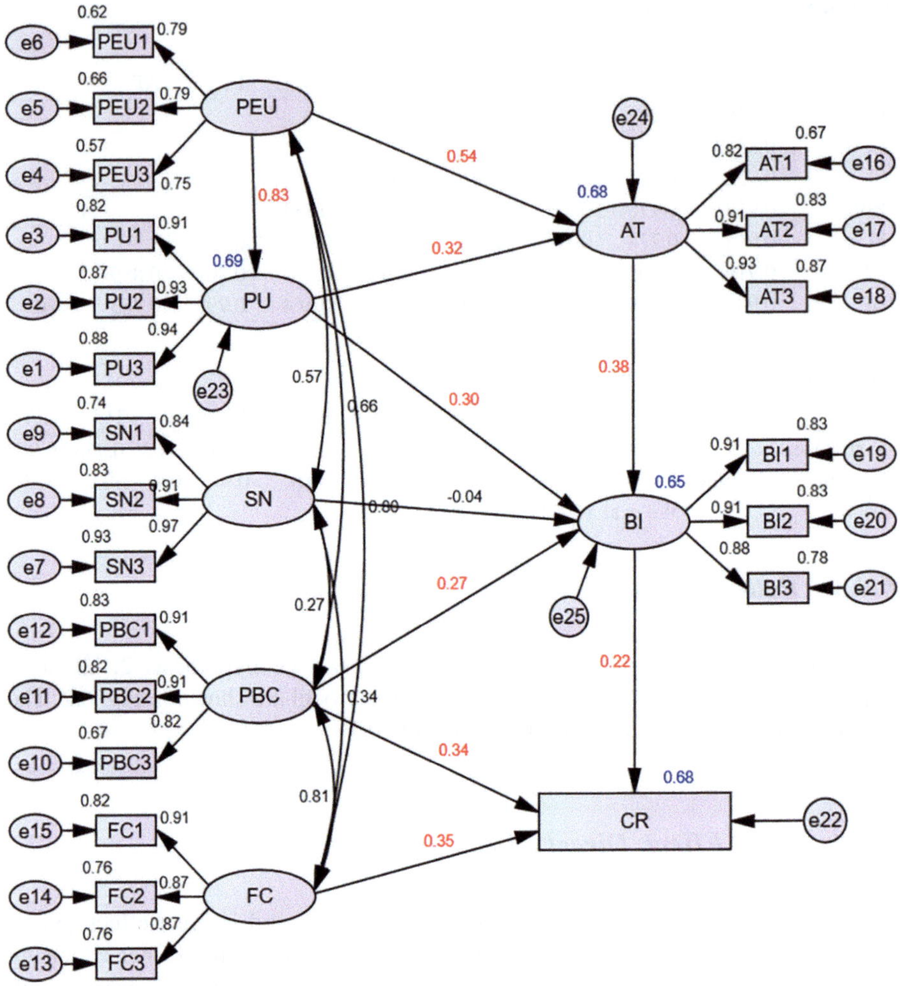

Fig. 6.2 Model analysis diagram

has an indirectly significant positive impact on the BI (0.553) through "AT→ BI (0.377)" and "PU→ BI (0.298)". Finally, PEU has an indirectly significant positive impact on CR (0.122) through "BI → CR (0.221)" (PEU has no direct effect on competition performance, but there is indirect effect: PEU → AT → BI → CR = 0.541 × 0.377 × 0.221 = 0.045; PEU → PU → AT → BI → CR = 0.834 × 0.321 × 0.377 × 0.221 = 0.022; PEU → PU → BI → CR = 0.034 × 0.298 × 0.221 = 0.055; indirect effect = 0.045 + 0.022 + 0.056 = 0.122; total effect = direct effect + indirect effect = 0 + 0.122 = 0.122; the calculation process of indirect effect and total effect of other paths are consistent with this). Davis, 1989 believes that individual experience in using new technologies significantly impacts PEU and PU While technology is simple to use, its usefulness will be improved accordingly. Notably, the total

Table 6.5 Hypothesis test results

Path	Hypothesis	Std.	S.E	C.R.	*p*	Verified results
BI <— AT	H1	0.377	0.077	5.634	***	Accepted
BI <— SN	H2	−0.043	0.052	−1.111	0.352	Rejected
BI <— PBC	H3	0.273	0.058	5.566	***	Accepted
CR <— BI	H4	0.211	0.048	4.622	***	Accepted
CR <— PBC	H5	0.344	0.068	5.946	***	Accepted
AT <— PU	H6	0.321	0.077	4.279	**	Accepted
AT <— PEU	H7	0.541	0.098	6.514	***	Accepted
PU <— PEU	H8	0.834	0.057	16.795	***	Accepted
BI <— PU	H9	0.298	0.072	4.935	***	Accepted
CR <— FC	H10	0.35	0.079	5.232	***	Accepted

Note: *** means $p < 0.001$, ** means $p < 0.01$

Table 6.6 Total effect, direct effect and indirect effect

		FC	PBC	SN	PEU	PU	AT	BI
PU	Total	–	–	–	0.834 ***	–	–	–
	Direct	–	–	–	0.834 ***	–	–	–
	Indirect	–	–	–	–	–	–	–
AT	Total	–	–	–	0.808 ***	0.321 **	–	–
	Direct	–	–	–	0.541 ***	0.321 **	–	–
	Indirect	–	–	–	0.267 **	–	–	–
BI	Total	–	0.273 ***	−0.043	0.553 ***	0.419 ***	0.377 ***	–
	Direct	–	0.273 ***	−0.043	–	0.298 ***	0.377 ***	–
	Indirect	–	–	–	0.553 ***	0.121 **	–	–
CR	Total	0.350 ***	0.404 ***	−0.009	0.122 ***	0.093 ***	0.083 ***	0.221 ***
	Direct	0.350 ***	0.344 ***	–	–	–	–	0.221 ***
	Indirect	–	0.060 ***	−0.009	0.122 ***	0.093 ***	0.083 ***	–

Note: *** means $p < 0.001$, ** means $p < 0.01$

effect of PEU and PU on BI is ranked first and second. Y Sun and S Cao (2021) believe that when college students feel that learning BIM technology is useful, they will take a positive attitude toward learning BIM technology. Zhao et al. (2022) verified that professionals' PEU and PU of BIM technology would significantly affect their attitude towards BIM technology. So, improving PEU and PU is an effective way to promote the willingness to use BIM technology and can further effectively improve students' performance in BIM competition. Therefore, in the teaching process of AEC majors, BIM modeling courses should be implanted earlier so that students can understand BIM technology earlier and realize the benefits of BIM technology to improve their professional ability and their acceptance of BIM technology. Moreover, students' PEU and PU of BIM technology should be improved through more practical operation links.

PBC has a significant positive impact on CR (0.404), and there are two influencing paths: PBC → CR, PBC → BI → CR. PBC has a directly and positively

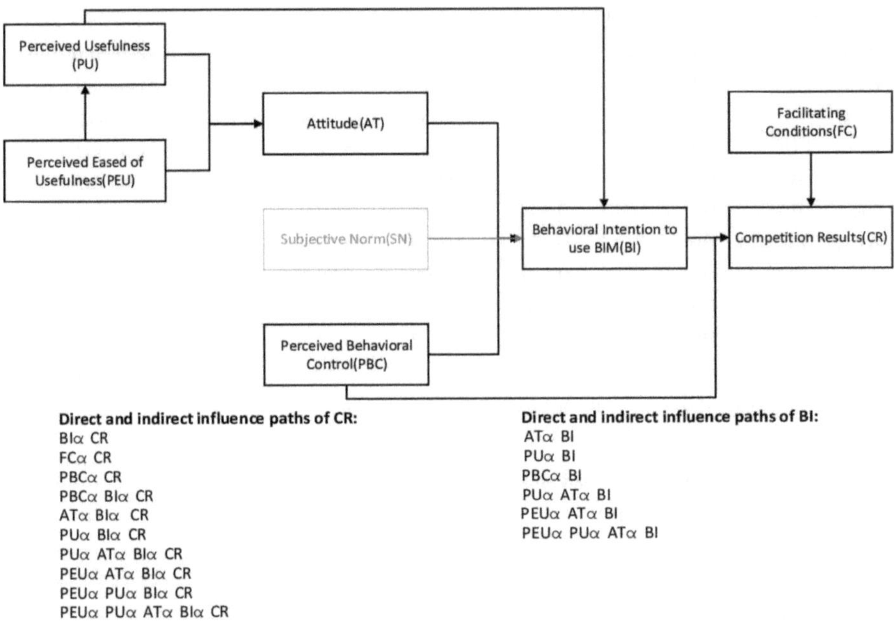

Fig. 6.3 Relationship diagram between direct and indirect effect variables

significant impact on BI (0.273) and CR (0.344). At the same time, the PBC has an indirectly and positively significant effect on CR (0.060) through "BI → CR (0.221)", which strengthens the total effect of PBC on CR. In conclusion, the overall effect of PBC on CR is the largest. This indicates that the stronger the confidence and belief of participants in completing the competition, the stronger the willingness to continue to learn BIM technology and engage in relevant positions in the future, which can indirectly improve the competitive performance of participants. Shi et al. (2019) believe that lack of confidence may lead students to give up participating in the competition, which is consistent with the results of this study. Meanwhile, if the participants have reasonable confidence and belief, they will keep a positive attitude in the competition process and learn BIM technology more actively. After an in-depth understanding of BIM technology, their willingness to use it in the future will be promoted (Howard et al., 2017). According to the result, improving PBC is an effective way to improve CR First, colleges should help students lay a good foundation in professional courses so that students could get familiar with the basic concept of BIM technology and software operation, which might help them to reduce the difficulty of participating in the competition. Second, colleges and enterprises can give students more opportunities to exercise their comprehensive ability to improve their self-confidence and give full play to their initiative in learning. This tool may help to improve the PBC of students.

FC, the second major factor affecting the CR, has a directly and positively significant impact on CR (0.350). Shi et al. (2019) pointed out that students are easy to be influenced by external factors, such as lack of competition information and inactive instructors. These influences hinder students from participating in competitions. Howard et al. (2017) also verified the positive relationship between external promoting conditions and students' personal use of BIM technology. It is consistent with the results of this study. This conclusion also reminds colleges to increase investment in BIM education and provide more suitable conditions to support students in learning BIM technology.

SN does not significantly influence BI, so there is no significant indirect influence on CR through "BI → CR". This result is different from the research conclusions of Howard et al. (2017) and Z. Wu et al. (2021), whose studies verified that individuals would be affected by surrounding people. The reason may be that their research is aimed at industry employees, and whether industry employees adopt BIM technology is more subject to the requirements of enterprises and industries (Howard et al., 2017). In contrast, the participants in this study are students, and their feelings and gains in the competition are more based on individual feelings. As a result, students are less affected by their surroundings.

6.6 Conclusions

6.6.1 Conclusions and Suggestion

In this study, participants of the eighth National BIM Graduation Design Innovation Competition were selected as the objects to explore factors that influence participants' performance and willingness to continue using BIM technology. A TAM-TPB-based driving factors model of the competition was built from the perspective of PEU, PU, AT, SN, PBC, FC, BI, and CR. The following conclusions are drawn through this study:

CR is directly and positively affected by BI, PBC, and FC, while it is indirectly and positively affected by AT, PBC, PU, and PEU. In addition, PBC has the most considerable total effect on CR.

BI is positively and directly significantly affected by AT, PBC, and PU, as well as positively and indirectly significantly affected by PU and PEU. Moreover, PEU has the largest total effect on BI.

To help ACE major students master BIM technology and complete the transition between school and industry, the following suggestions, which are combined with the current situation of BIM education, are put forward:

Increasing the publicity and promotion of BIM and related policies: In the current situation, students lack interest and initiative in BIM learning because of problems such as inadequate understanding. This study suggests that schools or enterprises carry out related lectures, activities, and competitions to preach BIM technology and relevant policies, and promote the development of BIM education through the "FC → CR" path. Moreover, professional course teachers can also have positive guidance for BIM technology to promote students' perceived usefulness. So that students would have a positive attitude to BIM technology and achieve good results in BIM competitions through the paths of "PU → AT → BI → CR" and "PU → BI → CR".

Strengthening the construction of supporting teaching facilities: At present, the supporting facilities and training sites of BIM teaching are not enough to support students in carrying out practical training. This survey shows that PEU has the largest total effect on BI, so it is important to improve students' PEU. Appropriate hardware and software support can help students learn BIM technology, thus improving academic PEU and helping to complete the competition. So, colleges should create good conditions, such as increasing the investment in teaching software and hardware, deepening school-enterprise cooperation, improving the basic practice teaching platform, building a special BIM training room, and completing BIM teaching software, for BIM teaching to motivate students to learn BIM and participate in competitions.

Building the integrated curriculum system of BIM technology: To improve the connection between professional courses at different stages, it is suggested to use the same project to teach in different courses. For example, use the same project to carry out teaching in courses such as architectural engineering drawing, architectural engineering budget estimate, REVIT, and others (Clevenger et al., 2007). Multi-stage training of the same project can effectively improve students' PEU and confidence in BIM learning and competition, and it could effectively improve competition performance and practical ability of students through the paths of "PEU → AT → BI → CR".

Strengthening the training of teachers: At present, most college teachers lack engineering practice experience. Therefore, colleges should organize industry lectures, training, and other activities. This pattern could help to strengthen the school-enterprise linkage and enrich the practical experience of teachers. People with rich practical experience in the industry can also be hired as guest lecturers to teach in case of insufficient teachers (Gao et al., 2021). Teachers with rich professional experience can effectively promote students' learning of BIM technology and their PEU, and thus realize a virtuous circle of "PEU → AT → BI → CR" path. In addition, they will provide professional guidance to contestants and improve students' performance in the competition through the "FC → CR" path.

6.6.2 Deficiencies and Further Research

Although the number of questionnaires is enough to complete the study, the questionnaire recovery rate is not ideal, which needs to be improved in future research. Moreover, the questionnaire was completed at one time. Participants completed the questionnaire according to their feelings during the whole process of the competition, and the process of attitude change during the competition was not taken into account.

In the following research, participants will be longitudinally tracked in different stages of the competition so that their competition situation and attitude change will be understood according to the tasks and progress of each stage. At the same time, the career development of participants who have participated in the competition would be followed up to understand their attitude changes after entering the workplace to provide a theoretical basis for improving the effect of BIM education and also help to complete the connection between school education and vocational positions.

References

Acquah, R., Eyiah, A. K., & Oteng, D. (2018). Acceptance of building information modelling: A survey of professionals in the construction industry in Ghana. *Journal of Information Technology in Construction, 23*, 75–91.

Ajzen, I. (1991). The theory of planned behavior. *Organizational Behavior and Human Decision Processes, 50*(2), 197–211. https://doi.org/10.1016/0749-5978(91)90020-T

Al-Ashmori, Y. Y., Othman, I., Rahmawati, Y., Amran, Y. M., Sabah, S. A., Rafindadi, A. D., & Mikić, M. (2020). BIM benefits and its influence on the BIM implementation in Malaysia. *Ain Shams Engineering Journal, 11*(4), 1013–1019.

Ao, Y., Liu, Y., Tan, L., Tan, L., Zhang, M., Feng, Q., … Martek, I. (2021). Factors driving BIM learning performance: Research on China's sixth national BIM graduation design innovation competition of colleges and universities. *Buildings, 11*(12). https://doi.org/10.3390/buildings11120616

Atabay, S., Pelin Gurgun, A., & Koc, K. (2020). Incorporating BIM and green building in engineering education: Assessment of a school building for LEED certification. *Practice Periodical on Structural Design and Construction, 25*(4), 04020040.

Autodesk. (2021). *What is BIM, building information modeling*. Retrieved 3 April 2022 from https://www.autodesk.eu/solutions/bim

Bosnjak, M., Ajzen, I., & Schmidt, P. (2020). The theory of planned behavior: Selected recent advances and applications. *Europe's Journal of Psychology, 16*(3), 352–356. https://doi.org/10.5964/ejop.v16i3.3107

Botero, G. G., Questier, F., Cincinnato, S., He, T., & Zhu, C. (2018). Acceptance and usage of mobile assisted language learning by higher education students. *Journal of Computing in Higher Education, 30*(3), 426–451. https://doi.org/10.1007/s12528-018-9177-1

Çakır, H. S., & Uzun, T. İ. (2020). Building information modelling in architectural education: Contribution of Bim in design process. *Turkish Online Journal of Design Art and Communication, 10*(4), 452–467.

Casasayas, O., Hosseini, M. R., Edwards, D., Shuchi, S., & Chowdhury, M. (2021). Integrating BIM in higher education programs: Barriers and remedial solutions in Australia. *Journal of Architectural Engineering, 27*(1), 05020010.

Clevenger, C., Glick, S. G., & Puerto, C. L. D. (2007). Interoperable learning leveraging building information modeling (BIM) in construction education. *International Journal of Construction Education and Research, 8*(2), 101–118. https://doi.org/10.1080/15578771.2011.647249

Davis, F. D. (1989). Perceived usefulness, perceived ease of use, and user acceptance of information technology. *MIS Quarterly, 13*, 319–340.

Davis, F. D., Bagozzi, R. P., & Warshaw, P. R. (1989). User acceptance of computer technology: A comparison of two theoretical models. *Management Science, 35*(8), 982–1003.

Elgewely, M. H., Nadim, W., ElKassed, A., Yehiah, M., Talaat, M. A., & Abdennadher, S. (2021). Immersive construction detailing education: building information modeling (BIM)–based virtual reality (VR). *Open House International, 46*(3), 359–375.

Fishbein, M., & Ajzen, I. (1975). *Belief, attitude, intention, and behavior: An introduction to theory and research.* Addison-Wesley.

Fornell, C., & Larcker, D. F. (1981). Evaluating structural equation models with unobservable variables and measurement error. *Journal of Marketing Research, 18*(1), 39.

Fu, L. (2018). Discussion on higher vocational colleges relying on vocational skills competition to improve the quality of personnel training. *Co-operative Economy & Science, 05*, 140–142.

Gao, Z., Wu, B., Shi, Y., & He, Y. (2021). Innovation and exploration of practical education mode of new engineering subject driven by BIM technology. In *2021 2nd international conference on education, knowledge and information management (ICEKIM).*

He, Q., Wang, G., Luo, L., Shi, Q., Xie, J., & Meng, X. (2017). Mapping the managerial areas of building information modeling (BIM) using scientometric analysis. *International Journal of Project Management, 35*(4), 670–685. https://doi.org/10.1016/j.ijproman.2016.08.001

Heng, L., & Hong-yu, L. (2020). Applications of BIM in construction engineering in China: A review. *E3S Web of Conferences, 143.*

Howard, R., Restrepo, L., & Chang, C.-Y. (2017). Addressing individual perceptions: An application of the unified theory of acceptance and use of technology to building information modelling. *International Journal of Project Management, 35*(2), 107–120. https://doi.org/10.1016/j.ijproman.2016.10.012

Jia, X. (2021). Research and education status analysis of building information Technology in China. *Real Estate Review, 09*, 29–33.

Jiang, J. (2020). Research on BIM second classroom teaching mode driven by subject competition. *Education Modernization, 7*(52), 70–73. https://doi.org/10.16541/j.cnki.2095-8420.2020.52.020

Lee, S., Yu, J., & Jeong, D. (2015). BIM acceptance model in construction organizations. *Journal of Management in Engineering, 31*(3). https://doi.org/10.1061/(asce)me.1943-5479.0000252

Li, J. (2021). Discussion on graduation design reform of engineering management specialty based on BIM. *Science & Technology Information, 19*(32), 112–114. https://doi.org/10.16661/j.cnki.1672-3791.2111-5042-6064

Lu, G., Chen, L., He, Q., & Yan, H. (2018). The evaluation of academic competition in universities: Plan, method and exploration. *China Higher Education Research, 2*(2), 63–68+74. https://doi.org/10.16298/j.cnki.1004-3667.2018.02.12

Ministry of Housing and Urban-Rural Development, P. (2022). *Science and technology development plan of housing and urban-rural construction in the 14th Five-Year Plan.* Retrieved 1 August from https://www.mohurd.gov.cn/gongkai/fdzdgknr/zfhcxjsbwj/202203/20220311_765107.html

Neto, I. L., Matsunaga, L. H., Machado, C. C., Günther, H., Hillesheim, D., Pimentel, C. E., … D'Orsi, E. (2020). Psychological determinants of walking in a Brazilian sample: An application of the theory of planned behavior. *Transportation Research Part F: Traffic Psychology and Behaviour, 73*, 391–398. https://doi.org/10.1016/j.trf.2020.07.002

Plakhotnik, M. S., Krylova, A. V., & Maslikova, A. D. (2020). Does participation in case competitions improve career decision-making self-efficacy of university students? *Education and Training, 62*(6), 659–675. https://doi.org/10.1108/et-01-2020-0021

Puolitaival, T., & Forsythe, P. (2016). Practical challenges of BIM education. *Structural Survey, 34*(4/5), 351–366. https://doi.org/10.1108/ss-12-2015-0053

Shi, M., Shi, L., & Ge, F. (2019). Factors influencing college students participating in discipline competition based on questionnaire survey. *Journal of Hubei University of Education, 36*(12), 78–82.

Su, S., & Zhang, F. (2021). A survey of factors influencing college students participating in discipline competition based on the questionnaire. *Fourth Industrial Review, 1*(2), 17–20.

Sun, Y., & Cao, S. (2021). Factors of willingness to learn BIM technology in college students based on technology acceptance model. *Project Management Technology, 19*, 70–76.

Venkatesh, V., & Zhang, X. (2014). Unified theory of acceptance and use of technology: U.S. Vs. China. *Journal of Global Information Technology Management, 13*(1), 5–27. https://doi.org/10.1080/1097198x.2010.10856507

Venkatesh, V., Morris, M. G., Davis, G. B., & Davis, F. D. (2003). User acceptance of information technology: Toward a unified view. *MIS Quarterly, 27*(3), 425–478. https://doi.org/10.2307/30036540

Wang, S. (2018). The applied undergraduate civil engineering major is based on BIM practica teaching research of design competition mechanism. *Information Recording Materials, 19*(1), 140–141. https://doi.org/10.16009/j.cnki.cn13-1295/tq.2018.01.091

Wang, Z., Guo, D., Wang, X., Zhang, B., & Wang, B. (2018). How does information publicity influence residents' behaviour intentions around e-waste recycling? *Resources, Conservation and Recycling, 133*, 1–9. https://doi.org/10.1016/j.resconrec.2018.01.014

Wang, Y., Pan, F., Huang, C., & Hu, Y. (2021). Research on "dual ability" innovative talent training mode of BIM assembly design competition and graduation project integration. In *2021 2nd international conference on education, knowledge and information management (ICEKIM).*

Wu, Z., Jiang, M., Li, H., Luo, X., & Li, X. (2021). Investigating the critical factors of professionals' BIM adoption behavior based on the theory of planned behavior. *International Journal of Environmental Research and Public Health, 18*(6), 3022. https://doi.org/10.3390/ijerph18063022

Xu, J., Li, B.-K., & Luo, S.-M. (2018). Practice and exploration on teaching reform of engineering project management course in universities based on BIM simulation technology. *Eurasia Journal of Mathematics, Science and Technology Education, 14*(5), 1827–1835.

Yang, Y., & Ha, M. (2018). Study on the influence factors of teaching quality of BIM in the mode of promoting teaching and learning by competition. *Journal of Hebei University of Engineering (Natural Science Edition), 35*, 106–109.

Zhang, J., Wu, W., & Li, H. (2018). Enhancing building information modeling competency among civil engineering and management students with team-based learning. *Journal of Professional Issues in Engineering Education and Practice, 144*(2), 05018001.

Zhao, X., Duan, X., Li, N., Luan, C., Zhu, H., & Zhang, X. (2020). Construction and application of BIM engineering technology research center. In *2020 international signal processing, communications and engineering management conference (ISPCEM).*

Zhao, Y., Sun, Y., Zhou, Q., Cui, C., & Liu, Y. (2022). How AEC professionals accept BIM technologies in China: A technology acceptance model perspective. In *Engineering, construction and architectural management* (ahead-of-print). https://doi.org/10.1108/ecam-04-2022-0308

Zhou, P., & Li, B. (2022). Influencing factors of secondary vocational students acceptance of blended learning: Based on the modified UTAUT mode. *Journal of Zaozhuang University, 39*(2), 111–118.

Zhou, Y., Yang, Y., & Yang, J.-B. (2019). Barriers to BIM implementation strategies in China. *Engineering Construction and Architectural Management, 26*(3), 554–574.

Chapter 7
Empowering Architecture, Engineering, and Construction Students Through Building Information Modeling Competitions: A Deep Dive into Behavioral Motivation

Abstract Purpose: Building Information Modeling (BIM) competitions are a beneficial approach to enhance BIM education, offering students practical experience in BIM application, including mastering workflows and technical tools. However, research exploring the individual perceptions influencing participation intentions and behaviors in BIM competitions is limited. Therefore, this study aims to investigate the factors affecting university students' behavioral intention and behavior in BIM competitions, providing theoretical support for BIM competitions and educational reform. Design/methodology/approach: This study employs the Structural Equation Modeling (SEM) based on the Unified Theory of Acceptance and Use of Technology (UTAUT) framework to analyze the factors influencing BIM competition participation among 970 Architecture, Engineering, and Construction (AEC) university students. Findings: The results of the study show that social influence, attitude, and self-efficacy play critical roles in shaping students' intentions to participate in BIM competitions. Furthermore, self-efficacy, facilitating conditions, and behavioral intention significantly influence students' actual engagement in such competitions. Surprisingly, effort expectancy negatively influences intentions, as less challenging tasks can lead students to perceive their participation as less impactful on their skills and learning, reducing their behavioral intention to participate. Originality/value: This research provides valuable insights into the effectiveness of BIM competitions in enhancing BIM education for AEC students. Extending the UTAUT model to include self-efficacy and attitude, provides a novel perspective for understanding students' intentions and behaviors regarding BIM competitions. The study's theoretical support proposes incorporating BIM competitions to augment BIM teaching methods and offers suggestions for advancing the efficacy of students' involvement in BIM competitions within higher education, thus contributing to educational reform in the AEC sector.

Keywords Building information modeling competition · Behavioral intention · Engineering education · Self-efficacy

Y. Ao et al., *Building Information Modeling (BIM) Adoption*, Resilient and Friendly Living Environments, https://doi.org/10.1007/978-981-95-3993-2_7

7.1 Introduction

The construction industry contributes significantly to world economic growth and urbanization, through its professional competence, advanced building technologies, continuous development, and innovative design concepts (Alaloul et al., 2021; Bock & Linner, 2015; Liang et al., 2022; Wang, 2014). These technologies and innovations not only improve the quality and efficiency of construction but also promote the practice of green building and sustainable development (Darko et al., 2017). With the emergence of the Internet of Things (IoT), industry 4.0, cloud computing, artificial intelligence, and other game-changing disruptions, the emergence of an 'intelligent building industry' has also opened up new opportunities (Fu et al., 2024; Jia et al., 2019). In particular, Building Information Modeling (BIM), remains at the forefront of this revolutionary movement, bringing about tremendous change to the way buildings are designed, constructed, maintained and operated (Abbasnejad et al., 2021).

BIM is an indispensable tool in architecture, engineering, and construction. It is a technology that computerizes the entire lifecycle process of engineering design, construction, and maintenance (Leite & Brooks, 2020). The use of BIM has revolutionized the construction industry by improving the efficiency of design and construction, facilitating collaboration and communication among all stakeholders, enabling centralized data management and analysis, and promoting the practice of design optimization and sustainability (Abbasnejad et al., 2021; Boshrabadi et al., 2021a). In addition, BIM is very helpful for AEC students because BIM software has been continuously developed and its application areas are expanding. In addition to the ability to create virtual models that reflect the reality of a building, BIM also helps to generate information about the entire life cycle of a building (Ding et al., 2015; Mandhar & Mandhar, 2013). As Kymmell (2008) points out, BIM is not limited to virtual reality because its models represent the physical characteristics of the actual building. Therefore, the characteristics of the learning environment of BIM are crucial for undergraduate architecture students (Abdirad & Dossick, 2016).

While BIM undeniably holds significant potential and importance in the construction sector, its educational landscape currently grapples with considerable challenges. (Li et al., 2020; Martek et al., 2021; Peng et al., 2022; Zhang et al., 2022). Education in BIM frequently lacks depth and breadth across institutions, and educators often face gaps in their understanding of BIM tools and concepts (Li et al., 2020). Struggling to keep pace with industry changes, educators are burdened by a disproportionate focus on software training, resulting in students possessing a unilateral, theoretical grasp of BIM without comprehension of its comprehensive workflow (Ahn & Kim, 2016; Casasayas et al., 2021; Tsai et al., 2019). Cheng (2006) pointed out that BIM education did not enhance learners' critical thinking abilities; instead, it led them to overly focus on software operational skills, neglecting deeper learning and thinking (Agirbas, 2020; Holzer, 2015). In higher education, professional skills are overly theorized, making it challenging to translate theoretical knowledge into practical application (Lynch et al., 2021). As a result, there are barriers to the transition of graduates from school to work in the labor market that is increasingly competitive (Nießen et al., 2022). Similarly, there is an apparent disconnect between education and industry in BIM education (Jin et al., 2018; Sotelino et al., 2020; Wang et al., 2020a), while the BIM education programs in colleges and universities do not fully meet the industry's need for BIM professionals. BIM education should not be limited to the application of tools but should enable AEC graduates to understand BIM workflows (Boshrabadi et al., 2021b; Sacks & Pikas, 2013). In addition, BIM education should improve the teaching of BIM concepts. BIM education should focus on practical outcomes, such that students fully internalize BIM concepts through the application of real projects. To address the disconnect between university education and industry practice, educators have begun to incorporate real-world projects into BIM education (Tsai et al., 2019). Such exposure to real-world projects enables students to apply BIM principles to solve practical problems while gaining insight into professional best practices (Zhang et al., 2022).

BIM competitions are often able to integrate real-world projects into teaching and learning, providing students with real-world project experience and application opportunities. Students can translate theoretical knowledge learned in the classroom into practical skills via real projects offered through BIM competitions (Alieh et al., n.d.; Ao et al., 2022; Zhao et al., 2022). At the same time, BIM competitions encourage innovation and improvement in teaching (Ao et al., 2022; Martek et al., 2021). Competition allows instructors to assess student learning outcomes and competencies, and to make course adjustments and improvements based on the competition results. In addition, BIM competitions foster collaboration and communication among students and improve teamwork skills (Ao et al., 2021). Participating students can also interact with industry professionals and judges, learn about the latest industry trends and practices, and enhance their understanding of BIM industry developments. BIM competitions have become an important platform for promoting BIM technology innovation and developing students' skills. These competitions not only stimulate students' creativity and encourage them to actively apply BIM technology to solve real architectural and engineering problems but also hone their collaboration and project management skills (Ao et al., 2022; Jin et al., 2018, 2019; Peng et al., 2022). Through competitions, students have the opportunity to translate their theoretical knowledge into practical skills and build a solid foundation for their future careers. However, students' behavioral intention to participate in BIM competitions and the extent to which they do so are influenced by a variety of factors. For example, social influence, facilitating conditions, as well as perceptions of the competition or themselves. An in-depth study of these factors is important to better understand the motivations and influences on students' behavior in participating in BIM competitions, to better stimulate their interest, increase their participation, and develop their overall skills.

Therefore, to fully leverage the significant role of BIM competitions in enhancing the training of BIM professionals and advancing the development of BIM vocational skills, researchers have started to examine students' participation behaviors in these competitions (Ao et al., 2021, 2022). However, there remains a scarcity of studies exploring the intentions and behaviors of BIM competition participants from the perspective of individual perceptions. As a result, this study aims to thoroughly investigate the factors that influence the behavioral intentions and actions of AEC university students in participating in BIM competitions, adopting an individual perspective. The goal is to provide specific guidance and decision-making support for competition organizers and relevant educational institutions, to foster the continuous advancement of these competitions, and to contribute to the training of professionals skilled in BIM technology and applications.

The chapter is structured as follows: Section 7.2 provides a review of existing literature on BIM education and related theoretical models. Section 7.3 presents the data and methodology employed in the study. The findings of the structural equation modelling are presented in Sect. 7.4. Section 7.5 offers a detailed analysis of these results. Section 7.6 summarizes the paper and provides recommendations.

7.2 Theoretical Background and Research Hypotheses

7.2.1 Theoretical Background

The Theory of Reasoned Action (TRA) and the Theory of Planned Behavior (TPB) are foundational models that have significantly contributed to our understanding of individual behavior, highlighting the influence of attitudes, subjective norms, and perceived behavioral control on behavioral intentions. Subsequent research led to the development of specialized models, such as the motivation model, the personal computer usage model, and the diffusion of innovations theory, each offering insights into behaviors within specific contexts. Despite their contributions, these models encountered limitations due to their narrow focus on certain factors. Addressing this gap, Venkatesh et al. (2003) introduced the Unified Theory of Acceptance and Use of Technology (UTAUT), amalgamating critical elements from preceding theories to enhance the model's relevance and precision across varied domains, thereby achieving greater predictive accuracy. The UTAUT model has found extensive application in education for investigating student acceptance of technologies or systems, as evidenced by studies like those of Agyei and Razi (2021) and Patil and Undale (2023). For instance, Mohan et al. (2020) and Altalhi (2020) applied the UTAUT model to assess MOOC adoption in higher education and to examine the dynamics between behavioral intentions and their determinants, while Chatterjee and Bhattacharjee (2020) employed UTAUT in studying the integration of artificial intelligence in higher education settings. Despite the broad validation and application of the UTAUT model, its acceptance framework is not without limitations. Some scholars advocate for integrating UTAUT with other relevant theories to boost its explanatory and predictive capabilities, as suggested by Al-Qeisi (2009), (Hu et al., 2020), and Isaac et al. (2019).

Among the four key variables in UTAUT, facilitating conditions are very similar to perceived behavioral control in the Theory of Planned Behavior (Godin & Kok, 1996). Perceived behavioral control refers to an individual's subjective evaluation and perception when performing a specific behavior. This includes a person's judgment on whether they have sufficient control to perform a specific action and an assessment of the required resources, skills, and knowledge. Ajzen (1991) believed that all factors that can influence behavior indirectly affect behavioral performance through behavioral intentions. Behavioral intentions are influenced by three related factors, one of which originates from perceived behavioral control. The subjective judgment of perceived behavioral control can be divided into external and internal factors (Ajzen, 2002). External factors refer to an individual's perception of external resources, while internal factors refer to an individual's level of internal confidence. This internal factor can also be known as self-efficacy (Bandura, 1986). Self-efficacy, as an individual's belief in their capability to execute a specific task, directly influences their behavioral intentions and behaviors. For example, an individual with high self-efficacy is more likely to adopt and effectively use new technology. Similarly, attitude—the individual's positive or negative evaluation of something—is also a key factor affecting the acceptance and use of technology. Positive attitudes can facilitate the acceptance and use of technology, while negative attitudes may hinder this process. Integrating self-efficacy and attitude into the UTAUT model can enhance the model's explanatory and predictive power regarding technology acceptance and usage behaviors.

Therefore, we propose extending UTAUT with self-efficacy and attitude to comprehensively understand AEC students' acceptance behavior towards BIM competitions. The remainder of this section will further elaborate on the model structure and propose related hypotheses.

7.2.2 Research Hypotheses

The theoretical framework model proposed in this study retains the four core variables of the UTAUT—performance expectancy, effort expectancy, social influence, and facilitating conditions—while further incorporating two key dimensions: attitude and self-efficacy. The following details each variable and hypothesis of the model.

Performance expectancy refers to the degree to which using technology is seen as effective in performing certain tasks. In this study, performance expectancy represents the extent to which individuals believe participating in BIM competitions helps improve their skills. Existing research indicates that performance expectancy is a primary determinant of behavioral intentions (Mugerwa et al., 2018; Rahi et al., 2019). Hence, the hypothesis is proposed:

H1: Students' perceived performance expectancy have a positive effect on behavioral intentions.

Effort expectancy refers to the anticipated effort required to use technology, representing in the context of BIM competition acceptance studies, college students' perceptions of the difficulty of participating in BIM competitions. Numerous studies confirm a positive correlation between effort expectancy and behavioral intentions. When students perceive BIM competitions as relatively easy, they may be more willing to participate. Therefore, this study hypothesizes:

H2: Students' perceived effort expectancy have a positive effect on behavioral intentions.

Social influence refers to the degree to which an individual perceives that important others (e.g., colleagues, friends, superiors) believe they should use the new technology. In environments with high social expectations, individuals are more likely to adopt new technology. In the context of BIM competitions, this includes students' perceived expectations and support from peers, teachers, and industry experts. It is believed that social influence positively affects students' behavioral intention to participate in BIM competitions because, in an environment of social support and encouragement, students are more likely to view participation as a commendable action. Thus, the hypothesis is proposed:

H3: Students' perceived social influence has a positive effect on behavioral intentions.

Furthermore, considering attitude is widely recognized as a key factor significantly influencing behavioral intentions (Botero et al., 2018; Davis et al., 1989), this study integrates attitude into the UTAUT. Here, an attitude refers to an individual's positive or negative evaluation of specific behavior, typically based on personal beliefs, emotions, and behavioral preferences. Therefore, this study proposes the following hypothesis:

H4: Students' attitudes toward BIM competitions have a positive effect on behavioral intentions;

Self-efficacy refers to an individual's expectation of their capability to perform a specific behavior in a particular context. Social learning theory considers self-efficacy a determinant of human behavior. A study by Agyei and Razi (2021) also indicates that self-efficacy has a positive effect on students' behavioral intentions when using Web-Based Instruction (WBI). Hence, the following hypothesis is proposed:

H5: Students' self-efficacy has a positive effect on their behavioral intentions to participate in BIM competitions.

In the context of BIM competitions, if students hold positive attitudes towards participation and believe they possess the capabilities needed to compete (i.e., have high self-efficacy), these factors will promote the formation of strong participation intentions. Subsequently, these behavioral intentions will positively influence actual behavior. However, students' actual behavior is not only driven by their behavioral intentions but is also affected by various facilitating factors or barriers encountered during the participation process. In this study, such influences are primarily manifested through facilitating conditions and self-efficacy. Facilitating conditions refer to the existence of organizational and technical infrastructure that individuals believe assists in performing a behavior. If students

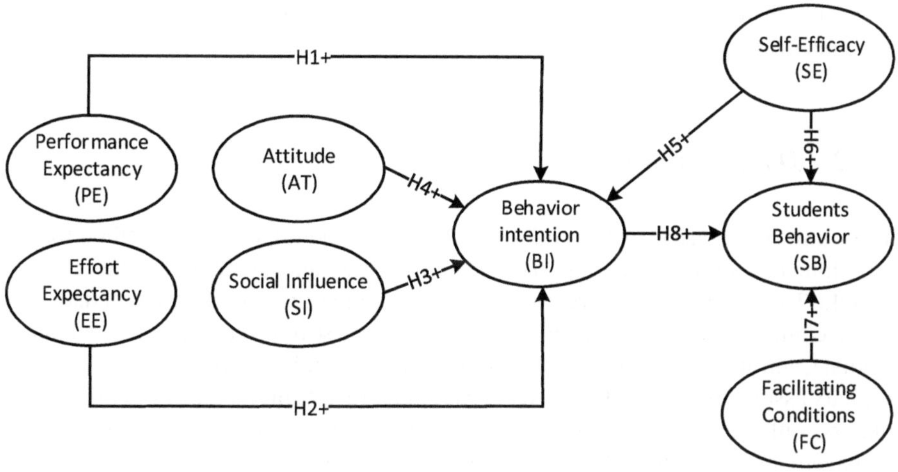

Fig. 7.1 Proposed research model. (Authors own work)

feel they have sufficient resources and support to participate in competitions, their likelihood of participation will increase.

Based on this theoretical framework and background analysis, the study proposes further hypotheses:

H6: Students' self-efficacy has a positive effect on their behavior;

H7: Students' perceived facilitating conditions have a positive effect on their behavior;

H8: Students' behavioral intentions have a positive effect on their behavior.

Based on the above assumptions, the theoretical framework model proposed in this study is shown in Fig. 7.1.

7.3 Materials and Methods

7.3.1 Questionnaire Design

For this study, a questionnaire was designed to test the intention and behavior of AEC students to participate in BIM competitions. It was based on existing established scales (Howard et al., 2017; Venkatesh et al., 2003), taking into account the actual situation of BIM competitions in China. The questionnaire consists of two main parts. The first record the demographic information of AEC students across major universities in China, while the second consists of 19 scaled questions that aim to measure students' intention and behavior to participate in BIM competitions. The scale was a five-point Likert scale, with 1–5 indicating " totally disagree," "basically disagree," "average," "basically agree," and "totally agree," respectively. The specific questions are listed in the Appendix.

7.3.2 Data Sampling

The study commenced in March 2022, utilizing a random sampling method for the questionnaire survey targeting specific universities and students as the research sample. To guarantee the representativeness and reliability of the research data, this study comprehensively included universities from seven geographical regions in China: Central, East, North, Northeast, Northwest, South, and Southwest. Given the primary association of BIM with AEC (Architecture, Engineering, and Construction) majors, only universities offering majors like civil engineering and construction management were selected. In line with the classification of Chinese undergraduate institutions, at least one university in each region was chosen, representing research-oriented, applied research-oriented, application-oriented, and private application-oriented categories. These targeted institutions were subsequently contacted. However, not every relevant university faculty was willing to participate in the research, and only a small number agreed to cooperate. Ultimately, 35 sample colleges and universities were identified, including 8 research-oriented universities, 10 applied research-oriented universities, 8 application-oriented universities, and 9 private application-oriented universities.

There were two stages in the data collection phase. The first was the pre-study phase, which lasted from March 20th to 26th, 2022. During this period, the research team conducted a questionnaire pilot at Chengdu University of Technology in Construction Management and subsequently improved the questionnaire based on the feedback. The formal research phase then ran from March 27th to April third, 2022, for a total of 6 days. The research team contacted faculty members in AEC-related disciplines and requested these faculty members to share a web-based questionnaire among students within their respective research schools, aiming to facilitate widespread participation in the study. A total of 1699 questionnaires were collected. However, 609 invalid questionnaires were eliminated as a result of screening for errors, inconsistencies, or other faults. Finally, 970 valid questionnaires were retrieved for analysis. The effective questionnaire return rate was 57.1%. The basic statistical information on the returned questionnaires is presented in Table 7.1.

Table 7.1 Descriptive statistics

Topic item	Option	Count heads	Percentage (%)
Gender	Male	402	41.44%
	Female	568	58.56%
Age	19 years or younger	49	5.05%
	19–21 years old	665	68.56%
	22–24 years old	250	25.77%
	Over 24 years old	6	0.62%
Grade	First year	181	18.66%
	Second year	318	32.78%
	Junior year	272	28.04%
	Senior year	199	20.52%
Specialized subject	Construction management	488	50.31%
	Construction costs	221	22.78%
	Civil engineering	85	8.76%
	Other AEC majors	176	18.14%
How do you know the BIM industry?	Understand	217	22.37%
	Neutral	552	56.91%
	Incomprehension	201	20.72%
Whether you have participated in BIM competitions or not?	Yes	180	18.56%
	No	790	81.44%
Whether the school offer BIM courses or not?	Yes	729	75.15%
	No	241	24.85%

7.4 Results

7.4.1 Testing for Reliability and Validity

This study used Cronbach's alpha coefficient for reliability testing (Bagozzi & Yi, 1988), which in general should be greater than 0.7 (Schmitt, 1996). The Cronbach's alpha coefficient for the data of this study was 0.915, so the reliability of this scale was quite good and passed the reliability test. In addition, the KMO value of the scale is 0.889, which is greater than 0.7 threshold, while the p-value is less than 0.05, which is significant. This means that the scale has passed the KMO and Bartlett's sphericity test and is suitable for factor analysis, which is conducted in the next step. The reliability test results are shown in Table 7.2.

A validated factor analysis of the data was then performed using AMOS 24.0. Convergent validity was jointly tested by standardized factor loadings, Average Variance Extracted (AVE), and Combined Reliability (CR) (Fornell & Larcker, 1981; Hair et al., 2012). The results showed that the factor loadings of all question items were all greater than 0.7, AVE was not less than 0.6, and CR was greater than 0.7, which passed the convergent validity test. Discriminant validity was tested

Table 7.2 Results of reliability test

Dimensions	Cronbach's α	KMO	Bartlett's P
PE	0.910		
EE	0.866		
SI	0.891		
AT	0.889		
SE	0.737		
FC	0.848		
BI	0.807		
Total	0.915	0.889	0

using the correlation coefficient between the AVE square root values and each variable. Good convergent validity was indicated since all AVE square root values were greater than their correlation coefficients. The correlation coefficients between the variables on this scale were smaller than the corresponding AVE square root values, with the single exception of the larger correlation coefficient between intention and attitude. Therefore, the discriminant validity of the scale was acceptable. The results of the specific validation factor analysis and the results of the validity test are presented in Tables 7.3 and 7.4.

Table 7.3 Results of confirmatory factor analysis

Measures		Factor loading	AVE	CR
PE	PE1	0.873	0.773	0.911
	PE2	0.874		
	PE3	0.89		
EE	EE1	0.736	0.693	0.871
	EE2	0.901		
	EE3	0.852		
SI	SI1	0.718	0.751	0.899
	SI2	0.921		
	SI3	0.943		
AT	AT1	0.828	0.735	0.893
	AT2	0.897		
	AT3	0.846		
SE	SE1	0.858	0.60	0.747
	SE2	0.681		
FC	FC1	0.857	0.736	0.848
	FC2	0.859		
BI	BI1	0.784	0.680	0.809
	BI2	0.863		

Table 7.4 Results of discriminant validity test

Variable	M	SD	PE	EE	SI	AT	SE	FC	BI
PE	4.42	0.01	**0.879**						
EE	3.11	0.16	0.292***	**0.833**					
SI	3.66	0.03	0.408***	0.397***	**0.867**				
AT	3.78	0.25	0.458***	0.295***	0.435***	**0.857**			
SE	4.32	0.25	0.375***	0.342***	0.405***	0.688***	**0.775**		
FC	3.44	0.03	0.533***	0.335***	0.505***	0.47***	0.414***	**0.858**	
BI	3.74	0.18	0.464***	0.237***	0.470***	0.859***	0.794***	0.456***	**0.824**

Note: Diagonal elements (the numbers in bold) are square root of average variance extracted (AVE), Off-diagonal elements are correlations, M mean, SD standard deviation, *** Means $p < 0.001$

7.4.2 *Model Fit Results*

The goodness of fit of the model is measured through five main metrics: CMIN/DF and Root Mean Square Error of Approximation (RMSEA), Comparative Fit Index (CFI), Tucker-Lewis Index (TLI), and Goodness of Fit Index (GFI). In general, the CMIN/DF of the structural equation model is optimal between 1 and 3, but less than 5 is acceptable if the sample size is larger. The RMSEA of this structural equation model is less than 0.08, and the CFI, GFI, and TLI are all greater than 0.9, indicating that the model fit is all relatively good (Z. Wu et al., 2021). The results of the fit are shown in Table 7.5.

7.4.3 *Hypothesis Testing Results*

In this study, AMOS24.0 was used to analyze the path of the theoretical model and verify the theoretical hypotheses, and the results of the analysis are shown in Table 7.6, and Fig. 7.2. The p-values for H3, H4, H5, H6, H7, and H8 were less than 0.05, and thus the original hypotheses were accepted. However, since the p-value of H1 is greater than 0.05, this original hypothesis is rejected. Similarly, since the path effect in H2 is negative, H2 is also rejected.

Each dimension had different effects on behavioral intention to participate and participation behavior. No statistically significant relationship was found between

Table 7.5 Results of model fitting

Model fitting index	Statistical value	Optimal standard value
CMIN/DF	4.163	<5
RMSEA	0.057	<0.08
GFI	0.942	>0.9
TLI	0.956	>0.9
CFI	0.966	>0.9

Table 7.6 Result of path analysis

Standard direct effect		Estimate	P	Test result
H1	BI←PE	0.045	0.145	Not supported
H2	BI←EE	−0.108	0.002	Not supported
H3	BI←SI	0.093	0.011	Support
H4	BI←AT	0.526	0.001	Support
H5	BI←SE	0.415	0.001	Support
H6	SB←SE	0.209	0.010	Support
H7	SB←FC	0.121	0.001	Support
H8	SB←BI	0.464	0.018	Support

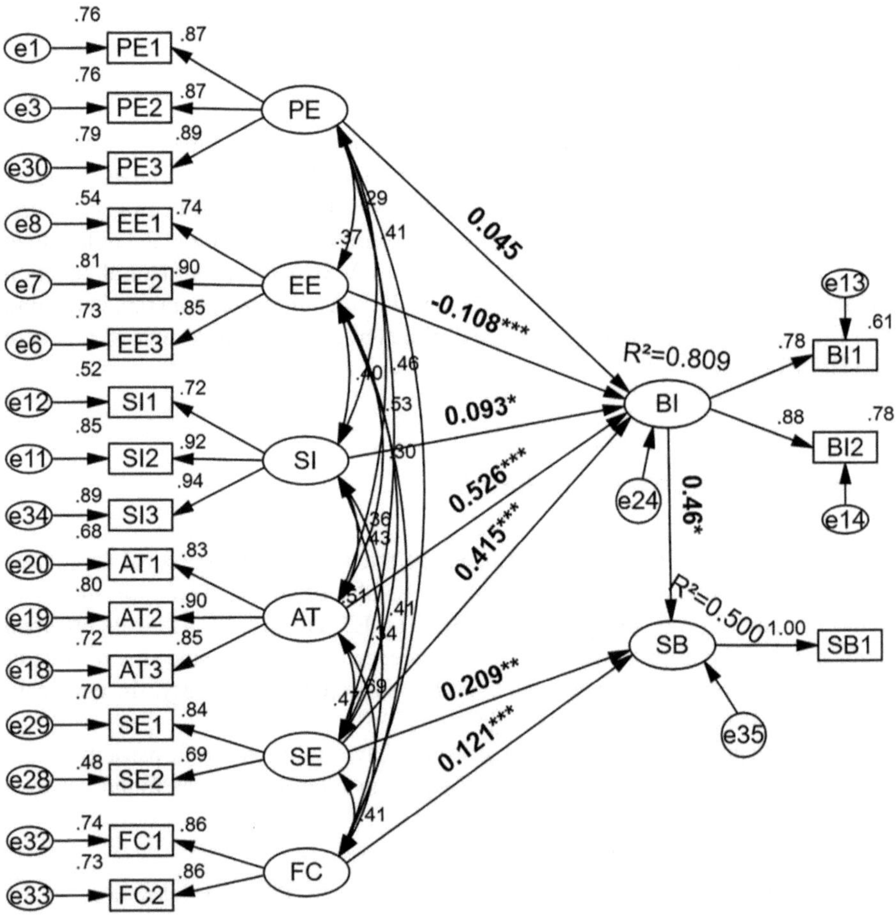

Fig. 7.2 Structural model analysis

Table 7.7 Intermediary effect

Standard indirect effect	β	P
PE→BI→SB	0.021	0.113
EE→BI→SB	−0.050	0.012
SI→BI→SB	0.043	0.011
AT→BI→SB	0.244.	0.017
SE→BI→SB	0.192	0.008

performance expectancy (β = 0.045, p = 0.145) and behavior intention. Social influence (β = 0.093, p = 0.002), attitude (β = 0.526, p = 0.001), and self-efficacy (β = 0.415, p = 0.001) all had a significant positive effect on students' behavioral intention to participate, with attitude having the greatest effect on behavioral intention, followed by self-efficacy. Self-efficacy (β = 0.209, p = 0.001), facilitating conditions (β = 0.121, p = 0.001), and behavioral intention (β = .464, p = 0.018) also had a significant positive effect on students' participation behavior. However, it should be noted that effort expectancy had a significant negative effect on behavioral intention (β = −0.108, p = 0.002), which contradicts the initial conjecture. The results of the squared multiple correlations showed that the extended UTAUT model explained 80.9% of students' behavioral intention to participate in BIM and 50.0% of their participation behavior.

To further verify the mechanism of the role of AEC students' behavioral intention and behavior to participate in BIM competitions, a bootstrap method was used to analyze the mediating effects under different pathways. As can be seen in Table 7.7, all mediating effects are significant except for the mediating effect of PE on SB. Attitude, Social Influence, and effort expectancy are fully mediated. Self-efficacy is partially mediated on behavior. The standardized indirect effects of attitude, social influence, and effort expectancy on behavior were 0.244, 0.043, and −0.050, respectively. The standardized indirect effect of self-efficacy on behavior was 0.192, the standardized direct effect was 0.209, and the standardized total effect was 0.402.

7.5 Discussion of the Results and Implications

7.5.1 Discussion of the Results

This study uses the UTAUT theory to explore the factors that influence the behavior of college students in AEC-related majors participating in BIM competitions. The relationships between performance expectancy, effort expectancy, social influence, facilitating conditions, self-efficacy, attitude, behavioral intention to participate, and behavior were examined.

1. Performance Expectancy Dimension

   According to Venkatesh et al. (2003), behavioral intentions are known to be influenced by performance expectations. However, in this study, the direct and indirect effects of performance expectations on participant intentions and behaviors were not significant. It is in college that students first participate in BIM competitions, and there is a large difference in participants' performance expectancy in regards to those competitions. Specifically, junior participants tend to be more active than senior participants, whose interest appears to wane. Therefore, the relationship between the effects of performance expectancy and intention to participate is not statistically significant.
2. Effort Expectancy Dimension

   In particular, the current study came to the surprising conclusion that effort expectancy have a significant negative effect on participation intentions and behavior. In layman's terms, this means that the easier people think it is to participate in the contest, the less likely they are to participate. Through further interviews with some of our students, the research team found that challenging tasks can stimulate interest and motivation, while simple tasks can lead participants to feel that participation makes a limited contribution to their skill development or learning experience, thus reducing their behavioral intention to participate. In addition, some students believe that although the BIM competition is relatively easy, the participants also need to spend a lot of time completing trivial tasks, such as modeling, optimizing the model, rendering, and so on. These trivial tasks can take a lot of time that may not be proportional to the reward. This is related to effort expectancy and may further indicate that students do not have sufficient knowledge about the effort expectancy of BIM competitions. Therefore, effort expectancy may negatively affect behavioral intention.
3. Social Influence Dimension

   This study also verified that Social Influence has a significant direct and indirect effect on college students' intention to participate in BIM. Studies have shown that young people are susceptible to peer influence (Choukas-Bradley et al., 2015), and when peers, teachers, and other people with influence over college students' behavior believe that he or she should participate in BIM competitions, this also increases their intention to participate. Given that BIM remains an emergent technology in the construction industry, BIM-related courses are required for AEC majors in many universities (Brokbals & Cadez, 2017; Sacks & Barak, 2010; Wang et al., 2020b). Thus, industry and education are working together to improve the teaching of BIM (Wu & Issa, 2014). College teachers are attentive to BIM technology, and so strongly encourage students to actively participate in such competitions with the imperative that BIM can improve their practical skills. Under such conditions, the social influence that students are subjected to will play a big role.
4. Attitude Dimension

   Through the validation of the BIM competition participation intention and behavior model, it was found that attitude ($\beta = 0.526$, $P = 0.001$) had the greatest predictive effect on college students' BIM participation intention, while attitude

($\beta = 0.244$, $P = 0.017$) had a significant indirect effect on behavior through intention. When college students have a positive attitude toward BIM competition, their intention to participate is stronger. It can be seen that when college students have more interest in BIM, their intention to learn is higher. This is similar to the findings of Mohan et al. (2020) who showed that when students find MOOCs more interesting, their intention to learn is stronger.

5. Facilitating Conditions and Self-Efficacy Dimensions

   The two dimensions of facilitating conditions and self-efficacy evolved from perceptual behavioral control in the Theory of Planned Behavior. These two factors interact to influence individual behavioral control. Facilitating conditions provide external environmental and resource support, while self-efficacy provides the individual's internal confidence and motivation. Therefore, this study's model, which added self-efficacy to the mix, also yielded interesting results. Facilitating conditions had a significant positive effect on participant behavior. This finding is consistent with previous research (Abbad, 2021; Hu et al., 2020). Similarly, self-efficacy had a significant positive effect on both intention and behavior to participate. Moreover, the effect of self-efficacy on intention and behavior was stronger. That is, the behavior of Chinese AEC university students to participate in BIM competitions is influenced by self-efficacy. When individuals have a high sense of self-efficacy, they believe that they can cope with the challenge and succeed, and therefore they are more likely to show positive intention and behavior toward participation. For example, a student who has strong confidence in his or her debating skills and knowledge on a debate competition theme will be more motivated to participate and try to do well in the competition. Therefore, this study found that the external environment had less of an effect on participant intentions and behaviors than the participant's self-confidence.

Moreover, although this study focuses on the factors influencing AEC students' behavior in participating in BIM competitions, it is equally crucial for preparing and developing students for their future careers. Participation in BIM competitions and mastery of BIM skills are valuable not only academically but also play a key role in professional practice. Therefore, the findings of this study can provide insights for educators and industry leaders, helping them design BIM courses and competitions that are closer to the realities of professional practice. In the real-world professional environment, the application of BIM skills extends far beyond the confines of academic learning. For example, in aspects such as construction project management, construction coordination, and facilities management, the application of BIM is becoming an industry standard. Thus, through BIM competitions, students can learn and apply these technologies and gain experiences similar to those in the actual work environment. This accumulation of experience is crucial for their smooth transition into professional roles. The results of this study also revealed a key point: students with higher self-efficacy and positive attitudes are more likely to show a behavioral intention and behavior to participate actively in BIM competitions. This finding is significant for career preparation, as these traits are often closely associated with professional success. The confidence and positive attitude cultivated by students in an academic environment could directly translate into their performance in future professional settings.

7.5.2 *Implications*

7.5.2.1 Managerial Implications

The body of findings reveal a way forward in further encouraging and supporting Chinese AEC students' education in BIM by way of participation in BIM competitions. Based on the results, this study offers the following recommendations to further enhance the effectiveness of student participation in BIM technology:

First, clearly communicate the goals and expectations of the competition to ensure that participants fully understand the goals, expectations, and values of the competition.

Second, enhance their confidence and self-efficacy by providing relevant skills training and resource support aimed at improving their perception of facilitating conditions, thereby increasing their intention and behavior to participate.

Third, understand participants' interests and goals and link the competition to their interests or career paths. Demonstrate what is interesting and meaningful about the competition by actively engaging them.

7.5.2.2 Theoretical Implications

This study applies the Unified Theory of Acceptance and Use of Technology (UTAUT) to the new context of participation behavior in BIM competitions, providing a fresh perspective for our understanding of technology acceptance theory. The findings reveal several key theoretical insights.

Firstly, the study found that effort expectancy has a negative impact on participation intentions, which contradicts the traditional belief that effort expectancy positively affects technology acceptance and participation intentions. This discovery suggests that the impact of effort expectancy may need to be re-evaluated in contexts with lower challenges. Secondly, social influence has a significant direct and indirect effect on participation intentions in educational and competitive environments. This emphasizes the importance of social factors in specific areas, such as education and technology competitions. Thirdly, the strong predictive effect of attitude on participation intentions highlights the central role of personal attitude in the technology acceptance model. This reveals the potential that individual intrinsic motivation and interest might surpass what was previously considered in the UTAUT framework. Additionally, the positive effects of self-efficacy and facilitating conditions on behavior offer new insights into the interaction between personal beliefs and external environmental support. This indicates that individuals with high confidence in their abilities are more likely to exhibit positive participation intentions and behaviors.

In summary, by extending the traditional UTAUT and demonstrating its adaptability and relevance in a specific context, this study provides new perspectives for the research field of technology acceptance theory. It calls for future research to more broadly consider the interplay between motivational factors, social dynamics, and the interaction of internal and external influences.

7.6 Conclusion

Based on the Unified Theory of Acceptance and Use of Technology, this study investigated the factors that shape the behavior of college students in AEC-related majors who participate in BIM competitions. The study examined the causal connections among performance expectancy, effort expectancy, social influence, facilitating conditions, self-efficacy, attitude, behavioral intention to participate, and behaviors, and drew the following conclusions:

Surprisingly, we found that effort expectancy negatively impacts students' behavioral intention to participate and actual participation behavior, suggesting a counterintuitive dynamic where perceived ease may deter engagement. This underlines the complexity of motivation in academic competitions. The significant positive role of social influence, primarily through encouragement from instructors and peers, emphasizes the importance of a supportive educational environment. Moreover, students' attitudes towards the BIM competition were the strongest predictors of their participation intention, indicating the pivotal role of personal interest and perceived value.

Facilitating conditions and self-efficacy emerged as critical factors enhancing participation intention and behavior, with self-efficacy showing a particularly strong positive effect. High self-efficacy students are more likely to embrace challenges and remain engaged, highlighting the need for educational strategies that build confidence and resilience.

By deeply exploring university students' intentions and behaviors towards participating in BIM competitions, this study not only demonstrates a profound understanding of the current topic but also stimulates further thought on future research directions. We realize that transforming experiences gained in the educational setting into applications in professional practice is a complex process, involving multifaceted factors at the individual, societal, and organizational levels. Therefore, future research should further investigate how students can translate their learning experiences from BIM competitions into skills and professional awareness in their career practices, including in-depth studies of cases where competition experiences are applied in actual work situations to provide richer insights.

Additionally, considering the smooth transition from educational experiences to professional practice is crucial for students' future success, future studies should also explore mechanisms that facilitate this process, such as continuing education programs, career development guidance, and practical opportunities in collaboration with the industry. These explorations will offer important perspectives for constructing a comprehensive framework to support students in effectively transferring and applying the knowledge and skills gained from competitions.

References

Abbad, M. M. M. (2021). Using the UTAUT model to understand students' usage of e-learning systems in developing countries. *Education and Information Technologies, 26*(6), 7205–7224. https://doi.org/10.1007/s10639-021-10573-5

Abbasnejad, B., Nepal, M. P., Ahankoob, A., Nasirian, A., & Drogemuller, R. (2021). Building information modelling (BIM) adoption and implementation enablers in AEC firms: A systematic literature review. *Architectural Engineering and Design Management, 17*(5–6), 411–433. https://doi.org/10.1080/17452007.2020.1793721

Abdirad, H., & Dossick, C. S. (2016). BIM curriculum design in architecture, engineering, and construction education: a systematic review. *Journal of Information Technology in Construction, 21*(17), 250–271.

Agirbas, A. (2020). Teaching construction sciences with the integration of BIM to undergraduate architecture students. *Frontiers of Architectural Research, 9*(4), 940–950.

Agyei, C., & Razi, Ö. (2021). The effect of extended UTAUT model on EFLs' adaptation to flipped classroom. *Education and Information Technologies, 27*(2), 1865–1882. https://doi.org/10.1007/s10639-021-10657-2

Ahn, E., & Kim, M. (2016). BIM awareness and acceptance by architecture students in Asia. *Journal of Asian Architecture and Building Engineering, 15*(3), 419–424. https://doi.org/10.3130/jaabe.15.419

Ajzen, I. (1991). The theory of planned behavior. *Organizational Behavior and Human Decision Processes, 50*(2), 197–211. https://doi.org/10.1016/0749-5978(91)90020-T

Ajzen, I. (2002). Perceived behavioral control, self-efficacy, locus of control, and the theory of planned behavior 1. *Journal of Applied Social Psychology, 32*(4), 665–683.

Alaloul, W. S., Musarat, M. A., Rabbani, M. B. A., Iqbal, Q., Maqsoom, A., & Farooq, W. (2021). Construction sector contribution to economic stability: Malaysian GDP distribution. *Sustainability, 13*(9), 5012.

Alieh, L. A. A., Hosseini, M. R., Martek, I., & Jupp, J. (n.d.). From BIM student to BIM professional: How work-ready are Australian Universities BIM graduates? *Industry4. 0 Applications for Full Lifecycle Integration of Buildings, 135.*

Al-Qeisi, K. I. (2009). *Analyzing the use of UTAUT model in explaining an online behaviour: Internet banking adoption*. Brunel University Brunel Business School PhD Theses.

Altalhi, M. (2020). Toward a model for acceptance of MOOCs in higher education: The modified UTAUT model for Saudi Arabia. *Education and Information Technologies, 26*(2), 1589–1605. https://doi.org/10.1007/s10639-020-10317-x

Ao, Y., Liu, Y., Tan, L., Tan, L., Zhang, M., Feng, Q., … Martek, I. (2021). Factors driving BIM learning performance: Research on China's sixth national BIM graduation design innovation competition of colleges and universities. *Buildings,* 11(12). https://doi.org/10.3390/buildings11120616.

Ao, Y., Peng, P., Li, J., Li, M., Bahmani, H., & Wang, T. (2022). What determines BIM competition results of undergraduate students in the architecture, Engineering and Construction Industry? *Behavioral Science, 12*(10). https://doi.org/10.3390/bs12100360

Bagozzi, R. R., & Yi, Y. (1988). On the evaluation of structural equation models. *Journal of the Academy of Marketing Science, 16*(1), 74–94.

Bandura, A. (1986). *Social foundations of thought and action*

Bock, T., & Linner, T. (2015). *Robot oriented design*. Cambridge University Press.

Boshrabadi, A. M., Oraee, M., Martek, I., & Hosseini, M. R. (2021a). BIM education assessment: Guidelines for making it authentic. In *BIM teaching and learning handbook* (pp. 294–304). Routledge.

Boshrabadi, A. M., Oraee, M., Martek, I., & Hosseini, M. R. (2021b). Incorporating collaborative problem solving (CPS) principles in BIM education. In *BIM teaching and learning handbook* (pp. 282–293). Routledge.

Botero, G. G., Questier, F., Cincinnato, S., He, T., & Zhu, C. (2018). Acceptance and usage of mobile assisted language learning by higher education students. *Journal of Computing in Higher Education, 30*(3), 426–451. https://doi.org/10.1007/s12528-018-9177-1

Brokbals, S., & Cadez, I. (2017). Academic teaching of BIM – Development – Status quo – Demand for action. *Bautechnik, 94*(12), 851–856. https://doi.org/10.1002/bate.201700100

Casasayas, O., Hosseini, M. R., Edwards, D. J., Shuchi, S., & Chowdhury, M. (2021). Integrating BIM in higher education programs: Barriers and remedial solutions in Australia. *Journal of Architectural Engineering, 27*(1), 05020010. https://doi.org/10.1061/(ASCE)AE.1943-5568.0000444

Chatterjee, S., & Bhattacharjee, K. K. (2020). Adoption of artificial intelligence in higher education: a quantitative analysis using structural equation modelling. *Education and Information Technologies, 25*(5), 3443–3463. https://doi.org/10.1007/s10639-020-10159-7

Cheng, R. (2006). Suggestions for an integrative education. In M. Broshar, N. Strong, & D. S. Friedman (Eds.), (pp. 1–10).

Choukas-Bradley, S., Giletta, M., Cohen, G. L., & Prinstein, M. J. (2015). Peer influence, peer status, and prosocial behavior: An experimental investigation of peer socialization of adolescents' intentions to volunteer. *Journal of Youth and Adolescence, 44*(12), 2197–2210. https://doi.org/10.1007/s10964-015-0373-2

Darko, A., Chan, A. P. C., Ameyaw, E. E., He, B.-J., & Olanipekun, A. O. (2017). Examining issues influencing green building technologies adoption: The United States green building experts' perspectives. *Energy and Buildings, 144*, 320–332. https://doi.org/10.1016/j.enbuild.2017.03.060

Davis, F. D., Bagozzi, R. P., & Warshaw, P. R. (1989). User acceptance of computer technology: A comparison of two theoretical models. *Management Science, 35*(8), 982–1003.

Ding, Z., Zuo, J., Wu, J., & Wang, J. (2015). Key factors for the BIM adoption by architects: A China study. *Engineering Construction and Architectural Management, 22*(6), 732–748.

Fornell, C., & Larcker, D. F. (1981). Evaluating structural equation models with unobservable variables and measurement error. *Journal of Marketing Research, 18*(1), 39.

Fu, H., Xia, Z., Tan, Y., & Guo, X. (2024). Influence of cues on the safety Hazard recognition of construction workers during safety training: Evidence from an eye-tracking experiment. *Journal of Civil Engineering Education, 150*(1), 04023009. https://doi.org/10.1061/JCEECD.EIENG-1882

Godin, G., & Kok, G. (1996). The theory of planned behavior: A review of its applications to health-related behaviors. *American Journal of Health Promotion, 11*(2), 87–98.

Hair, J. F., Sarstedt, M., Ringle, C. M., & Mena, J. A. (2012). An assessment of the use of partial least squares structural equation modeling in marketing research. *Journal of the Academy of Marketing Science, 3*(40), 414–433.

Holzer, D. (2015). BIM and parametric design in academia and practice: The changing context of knowledge acquisition and application in the digital age. *International Journal of Architectural Computing, 13*(1), 65–82.

Howard, R., Restrepo, L., & Chang, C.-Y. (2017). Addressing individual perceptions: An application of the unified theory of acceptance and use of technology to building information modelling. *International Journal of Project Management, 35*(2), 107–120. https://doi.org/10.1016/j.ijproman.2016.10.012

Hu, S., Laxman, K., & Lee, K. (2020). Exploring factors affecting academics' adoption of emerging mobile technologies-an extended UTAUT perspective. *Education and Information Technologies, 25*(5), 4615–4635. https://doi.org/10.1007/s10639-020-10171-x

Isaac, O., Abdullah, Z., Aldholay, A. H., & Ameen, A. A. (2019). Antecedents and outcomes of internet usage within organisations in Yemen: An extension of the unified theory of acceptance and use of technology (UTAUT) model. *Asia Pacific Management Review, 24*(4), 335–354.

Jia, M., Komeily, A., Wang, Y., & Srinivasan, R. S. (2019). Adopting internet of things for the development of smart buildings: A review of enabling technologies and applications. *Automation in Construction, 101*, 111–126. https://doi.org/10.1016/j.autcon.2019.01.023

Jin, R., Yang, T., Piroozfar, P., Kang, B.-G., Wanatowski, D., Hancock, C. M., & Tang, L. (2018). Project-based pedagogy in interdisciplinary building design adopting BIM. *Engineering Construction and Architectural Management, 25*(10), 1376–1397. https://doi.org/10.1108/ecam-07-2017-0119

Jin, R., Zou, P. X., Li, B., Piroozfar, P., & Painting, N. (2019). Comparisons of students' perceptions on BIM practice among Australia, China and UK. *Engineering Construction and Architectural Management, 26*(9), 1899–1923. https://doi.org/10.1108/ecam-07-2018-0275

Kymmell, W. (2008). *Building information modeling: Planning and managing construction projects with 4D CAD and simulations* (McGraw-Hill construction series). McGraw-Hill Education.

Leite, F., & Brooks, G. (2020). Integrating an architectural engineering undergraduate program with building information modeling. *Journal of Architectural Engineering, 26*(2). https://doi.org/10.1061/(asce)ae.1943-5568.0000389

Li, J., Afsari, K., Li, N., Peng, J., Wu, Z., & Cui, H. (2020). A review for presenting building information modeling education and research in China. *Journal of Cleaner Production, 259*, 120885. https://doi.org/10.1016/j.jclepro.2020.120885

Liang, L., Chen, M., & Lu, D. (2022). Revisiting the relationship between urbanization and economic development in China since the reform and opening-up. *Chinese Geographical Science, 32*(1), 1–15. https://doi.org/10.1007/s11769-022-1255-7

Lynch, M., Sage, T., Hitchcock, L. I., & Sage, M. (2021). A heutagogical approach for the assessment of internet communication technology (ICT) assignments in higher education. *International Journal of Educational Technology in Higher Education, 18*(1), 55. https://doi.org/10.1186/s41239-021-00290-x

Mandhar, M., & Mandhar, M. (2013). BIMing the architectural curricula: integrating building information modelling (BIM) in architectural education. *International Journal of Architecture, 1*(1), 1–20.

Martek, I., Wu, W., Oraee, M., & Hosseini, M. R. (2021). Educating the "T-shaped" BIM professional: Lessons from academia. In *BIM teaching and learning handbook* (pp. 247–261). Routledge.

Mohan, M. M., Upadhyaya, P., & Pillai, K. R. (2020). Intention and barriers to use MOOCs: An investigation among the post graduate students in India. *Education and Information Technologies, 25*(6), 5017–5031. https://doi.org/10.1007/s10639-020-10215-2

Mugerwa, S., Moya, M. B., & Mayoka Kituyi, G. (2018). Determinants of behavioral intention in adopting network monitoring system. *International Journal of Computer Applications Technology and Research, 07*, 139–157.

Nießen, D., Wicht, A., Schoon, I., & Lechner, C. M. (2022). "You can't always get what you want": Prevalence, magnitude, and predictors of the aspiration–attainment gap after the school-to-work transition. *Contemporary Educational Psychology, 71*. https://doi.org/10.1016/j.cedpsych.2022.102091

Patil, H., & Undale, S. (2023). Willingness of university students to continue using e-learning platforms after compelled adoption of technology: Test of an extended UTAUT model. *Education and Information Technologies*. https://doi.org/10.1007/s10639-023-11778-6

Peng, P., Ao, Y., Li, M., Wang, Y., Wang, T., & Bahmani, H. (2022). Building information modeling learning behavior of AEC undergraduate students in China. *Behavioral Science, 12*(8). https://doi.org/10.3390/bs12080269

Rahi, S., Mansour, M. M. O., Alghizzawi, M., & Alnaser, F. M. (2019). Integration of UTAUT model in internet banking adoption context: The mediating role of performance expectancy and effort expectancy. *Journal of Research in Interactive Marketing, 13*(3), 411–435.

Sacks, R., & Barak, R. (2010). Teaching building information modeling as an integral part of freshman year civil engineering education. *Journal of Professional Issues in Engineering Education and Practice, 136*(1), 30–38. https://doi.org/10.1061/(ASCE)EI.1943-5541.000000

Sacks, R., & Pikas, E. (2013). Building information modeling education for construction engineering and management. I: Industry requirements, state of the art, and gap analysis. *Journal of Construction Engineering and Management, 139*(11), 04013016. https://doi.org/10.1061/(ASCE)CO.1943-7862.0000759

Schmitt, N. (1996). Uses and abuses of coefficient alpha. *Psychological Assessment, 4*(8), 350–353.

Sotelino, E. D., Natividade, V., & Travassos do Carmo, C. S. (2020). Teaching BIM and its impact on young professionals. *Journal of Civil Engineering Education, 146*(4), 05020005. https://doi.org/10.1061/(asce)ei.2643-9115.0000019

Tsai, M.-H., Chen, K.-L., & Chang, Y.-L. (2019). Development of a project-based online course for BIM learning. *Sustainability, 11*(20). https://doi.org/10.3390/su11205772

Venkatesh, V., Morris, M. G., Davis, G. B., & Davis, F. D. (2003). User acceptance of information technology: Toward a unified view. *MIS Quarterly, 27*(3), 425–478. https://doi.org/10.2307/30036540

Wang, N. (2014). The role of the construction industry in China's sustainable urban development. *Habitat International, 44*, 442–450.

Wang, L., Huang, M., Zhang, X., Jin, R., & Yang, T. (2020a). Review of BIM adoption in the higher education of AEC disciplines. *Journal of Civil Engineering Education, 146*(3). https://doi.org/10.1061/(asce)ei.2643-9115.0000018

Wang, L., Yan, X., Fan, B., Jin, R., Yang, T., & Kapogiannis, G. (2020b). Incorporating BIM in the final semester undergraduate project of construction management — A case study in Fuzhou university. *KSCE Journal of Civil Engineering, 24*(8), 2403–2418. https://doi.org/10.1007/s12205-020-1971-4

Wu, W., & Issa, R. R. A. (2014). BIM education and recruiting: Survey-based comparative analysis of issues, perceptions, and collaboration opportunities. *Journal of Professional Issues in Engineering Education and Practice, 140*(2), 04013014. https://doi.org/10.1061/(asce)ei.1943-5541.0000186

Wu, Z., Jiang, M., Li, H., Luo, X., & Li, X. (2021). Investigating the critical factors of professionals' BIM adoption behavior based on the theory of planned behavior. *International Journal of Environmental Research and Public Health, 18*(6), 3022. https://doi.org/10.3390/ijerph18063022

Zhang, J., Zhang, Z., Philbin, S. P., Huijser, H., Wang, Q., & Jin, R. (2022). Toward next-generation engineering education: A case study of an engineering capstone project based on BIM technology in MEP systems. *Computer Applications in Engineering Education, 30*(1), 146–162. https://doi.org/10.1002/cae.22448

Zhao, Y., Sun, Y., Zhou, Q., Cui, C., & Liu, Y. (2022). How AEC professionals accept BIM technologies in China: a technology acceptance model perspective. *Engineering Construction and Architectural Management* (ahead-of-print). https://doi.org/10.1108/ecam-04-2022-0308

Chapter 8
Enhancing BIM Education for Future AEC Professionals: The Impact of Competition Experience on Student Perception and Career Choices

Abstract Building Information Modeling (BIM) has a profound impact on the future of the construction industry. However, the current BIM education in many universities worldwide fails to meet the industry's demand for BIM talent. The awareness and acceptance of BIM among university students have not been effectively studied. This research constructs a model of the driving factors for Architecture, Engineering, and Construction (AEC) students' focus on BIM applications based on the Theory of Planned Behavior (TPB) and the Technology Acceptance Model (TAM). It explores the different impacts of BIM competition experience on AEC students' acceptance of BIM positions. The results indicate that students with competition experience score higher in perceived ease of use, perceived usefulness, and attitude. Competition experience can positively influence students' choice of BIM positions. Therefore, it is recommended that universities and enterprises enhance students' BIM technical application abilities and career orientation through academic competitions and practical cooperation, thereby promoting the development of the BIM industry.

Keywords BIM competition · BIM adoption · Theory of planned behavior (TPB) · Technology acceptance model (TAM) · Difference analysis

8.1 Introduction

Over the past few decades, the architectural industry has not only achieved significant breakthroughs in technological innovation and design concepts (Elmualim & Gilder, 2014; Miettinen & Paavola, 2014), but has also played a crucial role in enhancing human living environments and fostering sustainable development in society (Sev, 2009). As global urbanization continues to

Y. Ao et al., *Building Information Modeling (BIM) Adoption*, Resilient and Friendly Living Environments, https://doi.org/10.1007/978-981-95-3993-2_8

escalate, the construction industry becomes increasingly vital in urban development. According to the World Economic Forum (WEF), the construction industry provides over 100 million jobs worldwide, and its value added represents 6% of global GDP (Agenda, 2016). The United Nations predicts that by 2030, approximately 61% of the global population will be urban (Grove, 2009), further spurring construction demand. Concurrently, the application of digital technology in building design and construction is on the rise. Building Information Modeling (BIM) technology, representing a significant transformation in the construction industry, profoundly influences its future trajectory (Ou et al., 2014; Sackey et al., 2015). However, despite the rapid development of BIM technology in the industry, there is a disconnect between the industry's demand for technical talent and university BIM-related courses (Peng et al., 2022). Effectively integrating BIM technology into higher education and cultivating future labor for the construction industry has become a challenge (Wu & Issa, 2014). While Architecture, Engineering, and Construction (AEC) majors in colleges and universities have been exploring effective BIM education methods and making some progress, there remains a gap in teaching methodologies that effectively bridge BIM education with industry practice (Jin et al., 2018). AEC students often lack comprehensive understanding of BIM workflows, indicating the need for further enhancement in BIM education (Tsai et al., 2019). To tackle the disconnect between university education and industry practice and to strengthen their alignment, the Chinese government, along with industry and academic institutions, have implemented a series of measures. Industry and academia worldwide, including in China, are continually seeking innovative collaborative approaches to advance professional development among students (Gu et al., 2021).

BIM-related digital construction presents a novel career path for college students, yet the educational process currently lacks effective incentives to motivate students towards BIM-related careers (Wu & Issa, 2014). While various BIM competitions exist within the industry, and recent studies have begun to examine student participation in these competitions (Ao et al., 2021; Ao et al., 2022; Peng et al., 2022), there remains a gap in research regarding the influence of competition experience on fostering student interest in BIM-related job positions. Therefore, this study aims to elucidate the role of competition experience in guiding students towards careers in the BIM sector. This paper conducted a survey among students enrolled in AEC majors, integrating the Theory of Planned Behavior (TPB) and the Technology Acceptance Model (TAM). This approach explores the impact of BIM competition experience on the inclination of college AEC graduates to pursue BIM-related positions. The findings of this study are intended to offer insights for the reformation and enhancement of BIM teaching practices within higher education in the AEC fields.

8.2 Literature Review

Based on existing researches, there are many problems in BIM education, and the effectiveness of BIM related competitions in promoting BIM education has not been effectively verified. The author will now explain the current situation of BIM education and BIM competitions from the following aspects in detail.

8.2.1 BIM Education

BIM has had a revolutionary impact on the development of the construction engineering field (Wang & Meng, 2021; Yang & Ha, 2018). However, with the application of BIM, the problem of the shortage of professional BIM technicians in the industry has emerged, and the lack of adequately trained professional BIM technicians is a major factor hindering the adoption of BIM in the construction industry (Li, 2021; Sotelino et al., 2020; Wu & Issa, 2014). In response to the demand for BIM talents in the construction industry, universities have taken the main responsibility of training BIM talents (Jin et al., 2018; Tsai et al., 2019). Currently, BIM education in universities is mainly divided into two aspects: management activities and technical activities in projects (Wang et al., 2020).The way of implementing BIM education is mainly divided into two types: integrating BIM into existing professional courses and offering separate BIM-related software courses (Elgewely et al., 2021; Tsai et al., 2019). BIM education in higher education has achieved some success, but there are some existing problems, such as change management challenges, limited course content, lack of professionals with engineering experience, too much focus on teaching software, students' one-sided knowledge of BIM, and lack of understanding of the complete BIM workflow (Ahn & Kim, 2016; Casasayas et al., 2021; Tsai et al., 2019). Tsai et al. (2019) concluded that BIM education should not be limited to the application of tools and should enable AEC graduates to understand the workflow of BIM. Therefore, they proposed a project-based online course for learning BIM. Ahn and Kim (2016) surveyed and studied Asian students' awareness and acceptance of BIM and BIM education, and found that students had some knowledge of BIM itself, but were unfamiliar with Industry Foundation Class (IFC), which is an open, vendor-neutral, international standard (ISO 16739–1:2018) aims to facilitate interoperability between various software platforms in the architectural and construction sector (Doe et al., 2022). Therefore, they believed that students' learning of BIM was incomplete and their knowledge of BIM was superficial and limited. Jin et al. (2019) compared students from three universities in Australia, the United Kingdom, and China and found that students in China had less positive attitudes toward the usefulness of BIM in the construction industry than students from the other two countries, which may be due to the industry environment in which BIM was applied in China.

Nießen et al. (2022) argued that in a competitive labor market, there are barriers to the transition of graduates between schooling and the workplace. Professional skills are theorized in higher education, which leads to difficulties in the transition between theoretical knowledge and practice for students. Similarly, there is a clear decoupling between education and industry in BIM for the construction profession (Jin et al., 2018; Sotelino et al., 2020; Wang et al., 2020). Fu et al. (2024) used eye-tracking technology to explore the impact of education cues and work experience on the identification of safety hazards during safety training for construction workers. They found that experienced workers performed better in all aspects of safety education models compared to novice workers. Therefore, companies tend to prefer students with experience when recruiting.

8.2.2 Project-Based Teaching of BIM

Comprehensive domestic and international literature from 2006–2019, the growth of BIM education research is in line with the development of the AEC industry, but the current BIM education does not meet the needs of the AEC industry, and there are problems such as uneven geographical development and uneven development of disciplines (Li et al., 2020). Although education in BIM has achieved some milestones with the promotion of BIM, there is still no widely recognized framework or system for BIM education and training (Sacks & Pikas, 2013; Wu et al., 2018). To address the disconnection between university education and industry practice, educators have begun to incorporate hands-on projects into BIM education. Integrating case studies into existing courses and leading students in BIM-related disciplinary competitions are more common ways to introduce hands-on projects into BIM education today. Van Nuland et al. (2015) confirmed that organized competition in the classroom can improve student performance through competitive activities among school students. Sayani et al. (2019) investigated their perceptions of project-based learning from the perspective of instructors and students, and both instructors and students agreed that the case teaching method has a positive effect on promoting student learning. The implementation of discipline competitions is mostly based on the cooperation between industry and universities. In an empirical study, Plakhotnik et al. (2020) found that participation in competitions had a positive effect on students' career decision-making, although the exact magnitude of the effect was not verified. Herrera-Limones et al. (2020) argued that participating in competitions takes both teachers and students out of their comfort zones, and that competitors grow professionally during the competition regardless of the outcome. Damnjanovic and Mijatovic (2017), on the other hand, suggested that participation in competitions transforms students from passive to active learning, while the competitive environment brings more learning resources to students. The positive effects of disciplinary competitions have been confirmed in studies, where the implementation of disciplinary competitions can expose students to real engineering projects, improve their core competencies, increase the recognition of students by companies, and increase their career maturity (Burke et al., 2013; Wankat, 2005; Wu, 2018), and educators should encourage students to participate more in competitions (Sadan, 2021). However, educators might rarely hear students' views from existing studies (Abernathy & Vineyard, 2001). On the other hand, the specific role and extent of the role of disciplinary competitions on students' careers has not been verified. It is important to further understand the impact of BIM disciplinary competitions on students' perceptions of BIM and BIM-related applications.

8.2.3 BIM Competition

With the increasing expansion of higher education, the relationship between higher education and employability has become increasingly tenuous; being highly educated does not mean being employable, as many college graduates are underemployed (Okay-Somerville & Scholarios, 2014). The traditional model of education no longer meets the demands of industry, so colleges and universities have begun to engage in pedagogical reforms to enhance the employability of college graduates and complete the transition from school to work (Donald et al., 2018). Disciplinary competitions in universities are based on classroom teaching, and they stimulate students' abilities such as theory and practice, independent thinking, and teamwork (Chen & Chen, 2020). Discipline competitions can cultivate students' practical and innovative abilities, promote the overall development of college students, enable students to train professional skills well, complete the transition from school to work, and greatly contribute to improving college students' employability. Therefore, disciplinary competitions are not only an important means of talent cultivation in colleges and universities, but also an important basis for employers to select talents, and have become an important part of practical education in school-enterprise cooperation (Donald et al., 2018; Plakhotnik et al., 2020). The National University BIM Graduation Design Innovation Competition has been successfully held for eight times, and the factors influencing the competition performance of participating college students have been studied to some extent (Ao et al., 2021; Ao et al., 2022). Nevertheless, the effect of the competition on students' movement toward BIM-related jobs has not yet been directly confirmed.

8.2.4 Application of Theoretical Models to BIM Adoption Behavior

The TPB, which proposes that Behavioral Intentions (BI) are influenced by three factors: Attitudes (AT), Subjective Norms, and Perceived Behavioral Control (PBC), has been widely applied in social and behavioral sciences, particularly in the fields of health sciences, environmental sciences, and educational research (Bosnjak et al.,

2020). The applicability of the TPB has also been validated in studies of BIM adoption behavior (Wu et al., 2021; Zhang et al., 2019). Z. Wu et al. (2021) adopted TPB empirical research to find that economic feasibility and government regulation were the most important factors influencing BIM adoption behavior among construction professionals in China. Zhang et al. (2019) applied TPB to study the efficiency of BIM adoption in sustainable construction and concluded that BIM behavioral intention, actual behavior, and behavioral attitude had a positive impact on sustainable construction.

The TAM was proposed in 1989 (Davis et al., 1989), which argues that people's use behavior towards new things is influenced by Perceived Ease of Use (PEU), Perceived Usefulness (PU), Attitude toward use, and Intention to use (Sun & Cao, 2021). The validity of the TAM as a way to predict and explain technology use has been tested in its application to the acceptance mechanism of BIM (Hong et al., 2019; Qin et al., 2020; Vigneshwar et al., 2022; Zhao et al., 2022). Zhao et al. (2022) and Qin et al. (2020) both studied the mechanisms and influencing factors affecting the application of BIM in the construction industry in relation to TAM and TOE (Technology-Organization-Environment), and they found that government incentives play an important role in the adoption of BIM in China. Hong et al. (2019) used the TAM to analyze the factors influencing the acceptance of BIM by construction personnel, and they found that behavioral control is one of the factors influencing the use of BIM, and they suggested that organizations provide incentives to encourage individuals to use BIM. Vigneshwar et al. (2022) used the TAM to analyze the drivers of BIM use among construction practitioners in India, and they suggested that both management and personnel capabilities can be strengthened to enhance the use of BIM.

TPB and TAM are two classical behavioral theory models, both developed from the Theory of Reasoned Action (TRA), and they have some homology. In previous studies, scholars have often integrated these two models to construct models applicable to various behaviors, such as studying tourists' willingness to consume green (Tang et al., 2022) and teachers' acceptance and use of educational portals (Pynoo et al., 2012). On the basis of existing researches, this study combined TPB with TAM to study the acceptance and behavior of AEC students towards BIM technology.

8.3 Methodology

8.3.1 Theoretical Framework

In order to explore the driving factors of BIM job acceptance and their influence relationships, the authors studied and constructed a model of the driving factors of AEC class students' attention to BIM jobs based on TPB and TAM (Ao et al., 2022; Peng et al., 2022). To further investigate the influence of BIM competition experience on AEC students' attention to BIM jobs, this study sets the variables of competition experience and constructs the theoretical model as shown in Fig. 8.1.

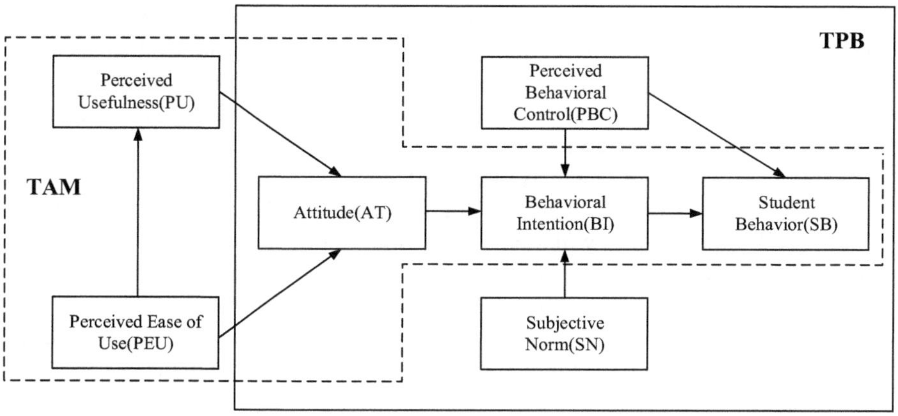

Fig. 8.1 Proposed structural model

8.3.2 Questionnaire Design

In this study, based on the theoretical model mentioned above and existing studies (Davis et al., 1989; Zhao et al., 2022), researchers have designed a questionnaire to assess AEC students' acceptance of BIM competition and BIM-related jobs, taking into account the actual situation of BIM education in China. The questionnaire is divided into two parts: the first part collects demographic information of the respondents, and the second part consists of 21 scale questions designed to measure the factors influencing students' acceptance of BIM competition and BIM-related jobs. The scale uses a five-point Likert scale, where 1–5 means "totally disagree", "basically disagree", "average", "basically agree", and "totally agree." The specific questions are listed in Appendix.

8.3.3 Data Collection

The data collection process for this study consisted of two phases. The first phase was conducted from March 21 to March 28, 2022. In this phase, the research team randomly selected a number of colleges and universities with AEC majors by region, and after contacting the faculty members of AEC majors in the research schools, the research team entrusted them to distribute the online questionnaires. Finally, 1699 questionnaires were collected, 609 invalid questionnaires were eliminated by polygraph questions (in which question students are asked to select a specified option), leaving 1090 valid questionnaires. In order to ensure the reliability of the questionnaire, this study further excluded samples with excessively short response times, and finally obtained 630 valid samples with no competition experience. The second phase was conducted from April 5 to May 28, 2022, and the questionnaire was administered to the participants of the eighth National University BIM Graduation Design Competition, who were undergraduate students majoring in the AEC category in major universities. The competition organizer, Glodon Company Limited sent out electronic questionnaires through the participants' online group, and finally 828 questionnaires were collected, of which 451 were valid (the questionnaire screening process was the same as the first stage).

Through the above two stages of data collection, the research team finally obtained 630 valid samples without participation experience and 451 with participation experience. Due to the fact that participating students only fill out the questionnaire as an obligation without any valid constraints, the number of students who complete the questionnaire seriously is limited, and the rate of collecting valid questionnaires is relatively low. Table 8.1 shows the descriptive statistics of the valid questionnaires.

8.4 Result

8.4.1 Reliability and Validity Tests

Reliability and validity tests were initially performed following data collection. Firstly, reliability analysis of the data was conducted using Cronbach's alpha coefficient in SPSS 26.0, then correlation and bias correlation between variables were checked using KMO test and Bartlett's sphericity test. For the reliability analysis, the Cronbach's alpha coefficient for the participating group was 0.958 and the alpha coefficient for the non-participating group was 0.879, both higher than 0.7. The

Table 8.1 Results of descriptive statistics

	Non-participating group		Participating group	
Variable	Frequency	Percentage	Frequency	Percentage
Total	630	100%	451	100%
Gender				
Male	260	41%	251	56%
Female	370	59%	200	44%
Grade				
First year	124	20%	13	3%
Second year	196	31%	116	26%
Junior year	184	29%	197	44%
Senior year	126	20%	125	28%
Major				
Construction management	331	53%	125	28%
Engineering cost	151	24%	146	32%
Civil engineering	47	7%	100	22%
Other	101	16%	77	17%

reliability passed the test Schmitt (1996). Meanwhile, the KMO values of both data groups were greater than 0.8 and Sig. <0.05 (i.e., p-value <0.05), which were suitable for factor analysis and could be tested in the next step. The results of the tests are presented in Table 8.2.

The confirmatory factor analysis was conducted using AMOS24.0, and each variable in the two groups of data was analyzed respectively. The factor loadings for each question item were greater than 0.6. In addition, convergent validity tests were conducted using two indicators, average variance extraction (AVE) and combined reliability (CR). Good convergent validity was indicated when AVE > 0.5 and when CR > 0.7 (Fornell & Larcker, 1981). The results showed that all AVEs were greater than 0.5 and all CRs were greater than 0.7, which passed the convergent validity test. The square root of the AVE and the correlation coefficient were used to test the discriminant validity. The square root of AVE was greater than the correlation coefficient on the horizontal and vertical axes, indicating good discriminant validity (Hair et al., 2012). The results of the validity tests are presented in Tables 8.3 and 8.4.

8.4.2 Model Fitting Results

In this study, the discrepancy divided by degree of freedom (CMIN/DF) and Root Mean Square Error of Approximation (RMSEA), Comparative Fit Index (CFI), Tucker-Lewis Index (TLI), and Goodness of Fit Index (GFI) were used as indicators to examine the model fit. The model fit results are shown in Table 8.5. The CMIN/DF of both samples in this model are between 1 and 3, the RMSEA is less than 0.08, and the CFI, GFI, and TLI are all greater than 0.9, indicating that the model fit is good (Wu et al., 2021).

Table 8.2 Results of reliability test

Dimensions	Non-participating group			Participating group		
	Cronbach's α	KMO	Bartlett's P	Cronbach's α	KMO	Bartlett's P
PEU	0.910			0.865		
PU	0.877			0.963		
SN	0.934			0.930		
PBC	0.847			0.865		
AT	0.839			0.915		
BI	0.793			0.921		
SB	0.853			0.924		
Total	0.879	0.839	0.000	0.958	0.944	0.000

Table 8.3 Results of confirmatory factor analysis

Measures		Non-participating group			Participating group		
		Factor loading	CR	AVE	Factor loading	CR	AVE
PEU	PEU1	0.905	0.913	0.778	0.795	0.867	0.685
	PEU2	0.941			0.846		
	PEU3	0.793			0.841		
PU	PU1	0.764	0.880	0.647	0.912	0.963	0.868
	PU2	0.829			0.937		
	PU3	0.789			0.934		
	PU4	0.833			0.944		
SN	SN1	0.847	0.935	0.828	0.842	0.933	0.823
	SN2	0.945			0.912		
	SN3	0.935			0.963		
PBC	PBC1	0.920	0.853	0.662	0.880	0.872	0.694
	PBC2	0.767			0.808		
	PBC3	0.742			0.810		
AT	AT1	0.842	0.841	0.638	0.817	0.919	0.791
	AT2	0.821			0.915		
	AT3	0.729			0.932		
BI	BI1	0.769	0.801	0.573	0.928	0.921	0.795
	BI2	0.801			0.893		
	BI3	0.698			0.852		
SB	SB1	0.826	0.855	0.747	0.871	0.929	0.868
	SB2	0.901			0.989		

8.4.3 Model Test Results

In this study, AMOS 24.0 was used to test the theoretical model, and the model test results are shown in Table 8.6. Based on Table 8.6, researchers plotted the results of the structural equation model as shown in Fig. 8.2, where the solid lines represent the paths and coefficients of the participating groups and the dashed line represents the paths and coefficients of the non-participating groups. No path coefficients are reported to indicate that there is no statistically significant relationship between the variables.

8.5 Discussion

The purpose of this study was to investigate whether the presence or absence of participation experience has an effect on the acceptance of BIM-related jobs among college students in AEC majors using the TAP-TAM model. Overall, the magnitude of the path coefficients as well as the significance differed between students with

Table 8.4 Results of discriminant validity test

Non-participating group			Correlations						
Variable	M	SD	PEU	PU	SN	PBC	AT	BI	SB
PEU	3.730	0.140	**0.882**						
PU	4.508	0.017	0.475***	**0.804**					
SN	3.623	0.059	0.196***	0.328***	**0.910**				
PBC	3.423	0.086	0.254***	0.250***	0.263***	**0.813**			
AT	3.733	0.172	0.259***	0.347***	0.157***	0.291***	**0.799**		
BI	3.670	0.295	0.209***	0.390***	0.341***	0.245***	0.647***	**0.757**	
SB	2.885	0.007	0.171***	0.205***	0.291***	0.249***	0.362***	0.578***	**0.864**
Participating group			Correlations						
Variable	M	SD	PEU	PU	SN	PBC	AT	BI	SB
PEU	4.053	0.115	**0.828**						
PU	4.210	0.062	0.815***	**0.932**					
SN	4.050	0.036	0.599***	0.52***	**0.907**				
PBC	3.733	0.057	0.588***	0.615***	0.326***	**0.889**			
AT	3.887	0.055	0.672***	0.769***	0.371***	0.747***	**0.833**		
BI	4.007	0.070	0.646***	0.701***	0.309***	0.721***	0.759***	**0.892**	
SB	3.855	0.035	0.636***	0.632***	0.327***	0.653***	0.666***	0.842***	**0.932**

Note: Diagonal elements (the numbers in bold) are square root of average variance extracted (AVE), Off-diagonal elements are correlations, M mean, SD standard deviation, *** Means $p < 0.001$

Table 8.5 Results of model fitting

Model fitting index	Non-participating group	Participating group	Recommended value
	Statistical value	Statistical value	
CMIN/DF	2.476	2.572	1 ~ 3
RMSEA	0.048	0.059	<0.08
GFI	0.937	0.912	>0.9
TLI	0.961	0.965	>0.9
CFI	0.967	0.970	>0.9

and without participation experience, i.e., the presence or absence of participation experience affects students' acceptance of BIM and BIM-related jobs.

First, the presence or absence of participation experience caused a difference in the effect of perceived ease of use and perceived usefulness on student behavior. As shown in Table 8.6 and Fig. 8.2, the paths of perceived ease of use and perceived usefulness on student behavior can be divided into PEU → PU → AT → BI → SB, PEU → AT → BI → SB, and PU → AT → BI → SB. In the path of PEU → PU → AT → BI → SB, the path coefficient was greater for the participating group than for the non-participating group. This indicates that participation in BIM competitions has a positive effect on improving students' perceived ease of use and

Table 8.6 Results of structural equation model test

	Non-participating group		Participating group	
Standard Direct Effect	β	*P*	β	*P*
PEU → PU	0.480	***	0.841	***
PEU → AT	0.119	*0.038	0.240	*0.012
PU → AT	0.301	***	0.577	***
AT → BI	0.614	***	0.543	***
SN → BI	0.261	***	0.005	0.907
PBC → BI	0.021	0.657	0.371	***
BI → SB	0.551	***	0.765	***
PBC → SB	0.114	*0.030	0.107	0.062
	Non-participating group		**Participating group**	
Standard indirect effect	β	*P*	β	*P*
PEU → AT → BI	0.073	*0.034	0.13	***
PEU → AT → BI → SB	0.04	*0.032	0.1	***
PEU → PU → AT	0.145	***	0.485	***
PEU → PU → AT → BI	0.089	***	0.263	***
PEU → PU → AT → BI → SB	0.049	***	0.202	***
PU → AT → BI	0.185	***	0.313	***
PU → AT → BI → SB	0.102	***	0.24	***
SN → BI → SB	0.144	***	0.004	0.907
PBC → BI → SB	0.012	0.634	0.284	***

Note:***: $p < 0.001$; *: $p < 0.05$

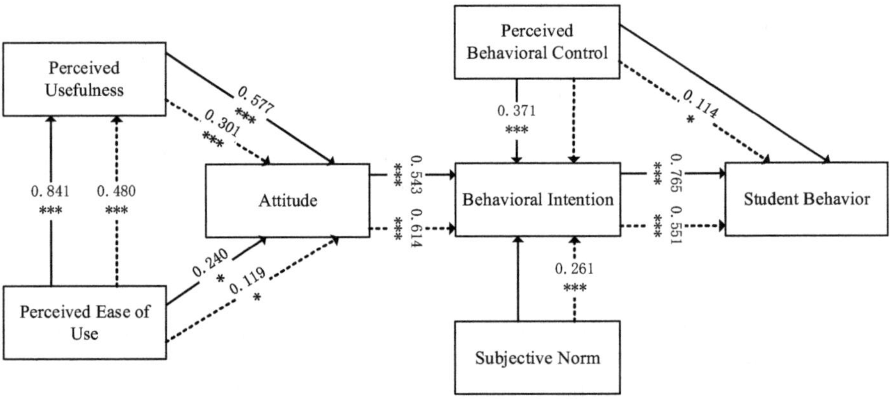

Fig. 8.2 Results of structural model

perceived usefulness of BIM, and this effect may further contribute to attitude change. It may be because the competition experience and BIM training of the participating group made them more proficient in BIM, and their perceptions of the ease of use and usefulness of BIM were more ac4.curate and deeper. Students with competition experience are more inclined to associate the ease of use with the usefulness of BIM, and are more willing to accept and use BIM. At the same time, they may pay more attention to the ease of use and usefulness of BIM in practical application scenarios because of the competition experience, and thus pay more attention to the perception of this aspect and translate it into a more positive attitude. The non-participating group was less skilled and confident in the use and application of BIM due to their lack of experience in related competitions and training, and thus were relatively weaker in terms of perceived ease of use and had a less positive attitude toward BIM than the participating group. However, it is worth noting that the perceived usefulness of the non-participating group (M = 4.508) is stronger than that of the participating group (M = 4.210), which may be due to the limited knowledge of BIM in the non-participating group, which is easily attracted by others' views and new technologies, and lacks a comprehensive consideration of the practical applications and limitations of BIM. In the AT→BI→SB pathway, the influence of attitude on willingness was stronger in the non-participating group, but the influence of attitude on behavior was higher in the participating group. This may be due to the fact that students in the participating group are more practice and action-oriented. They are more capable of translating their attitudes into practical actions, thus increasing their willingness to behave. It is also indicated that participation in BIM competitions can improve students' action and acceptance of BIM-related jobs. At the same time, the competition experience provides students more opportunities to practice and master their skills, which also helps them better cope with challenges in their future careers. Therefore, participation in BIM competitions has a positive effect on the overall quality improvement of students, and it is worthwhile for more students to be engaged.

Second, there was a difference between the presence and absence of participation experience in the effect of subjective norms on behavioral intentions. There was no statistically significant relationship between the participation group in the SN → BI pathway. This may be because students with participation experience already have higher awareness and acceptance of BIM, and they may have gained in-depth understanding and practice opportunities through participation in the competition, so subjective norms have less influence on them. At the same time, students with participation experience may be more inclined to actively learn BIM, with stronger interest and higher commitment to BIM, thus reducing the influence of subjective norms. Students without participation experience, on the other hand, need more guidance and cognitive construction in this regard.

Finally, the presence or absence of participation experience differed in terms of perceived behavioral control. In the PBC → BI pathway, the effect of perceived behavioral control on behavioral intentions was stronger in the participating group than in the non-participating group. This may be because students with participation experience have a better understanding of BIM, and they value the level of resources available to them more highly when engaging in behavior. In addition, students with competition experience tend to be more proactive students, who may be more active in learning BIM-related knowledge at university and gaining experience in participating in competitions, and thus behave more positively in terms of perceived behavioral control. This shows that participation in competitions improves students' mastery of technology and practical experience, thus increasing their confidence and willingness to use technology. However, the direct effect of perceived behavioral control on behavior ($\beta = 0.107$, $p = 0.062$) in the participant group was not significant, but the overall effect of the path PBC → BI → SB ($\beta = 0.284$, $p < 0.001$) was made significant after the indirect effect through BI. Therefore, the indirect effect of perceived behavioral control on students' behaviors was the largest, indicating that the perceived behavioral control of the competition group mainly influenced behaviors through behavioral intentions. That is, the more resources or the more solid BIM knowledge students with participation experience have, the stronger the students' willingness to continue learning BIM and working in related positions in the future, and they will take appropriate actions.

The results of this study suggest that educators need to focus on cultivating students' practical operation and skill control on the one hand, whilst cultivating students' cognitive and attitudinal aspects. Participation in BIM competitions to improve their cognition of BIM when studying could enhance students' attitudes towards BIM and BIM-related jobs.

8.6 Conclusion

8.6.1 Findings

This study explores the differential effects of BIM competitions on the acceptance of BIM-related jobs among AEC majors college students based on the TPB-TAM integration model by combing the relevant literature and starting from individual perception factors. The study shows that students with participation experience have significantly more positive attitudes toward continuing to adopt BIM and entering the BIM industry than students without participation experience. This shows that the BIM competition experience encourages students to pay more attention to BIM and its related jobs. Further analysis revealed that students who had participated were more likely to agree with the importance of BIM and were more likely to use BIM in their future work. This may be due to the fact that BIM competitions can provide a practical platform that enables students to better understand and master the practical application of BIM. At the same time, participation in the competition can build students' confidence and professionalism, making them more likely to pursue BIM-related careers. In conclusion, BIM competitions can encourage students to pay more attention to BIM and BIM-related jobs, and provide an effective way for universities and industries to train more BIM talents.

8.6.2 Development Strategies

The study further provides the following suggestions to provide the theoretical basis for BIM education teaching and competition, help ACE students to better master BIM, and better complete the transition from school education to industry practice. To provide more accurate talent training and recruitment goals for universities and industries, to export BIM talents for the construction industry in a sustainable and stable way, and to promote the healthy development of the BIM applications in AEC industry.

1. Universities and enterprises should hold relevant discipline competitions to cultivate students' innovation consciousness and cooperation spirit, and improve their innovative design ability, practical ability and overall quality.
2. Universities can take appropriate incentive measures, such as strengthening the organization and publicity of BIM competitions, and providing more support and rewards for students who participate in the competitions, so as to encourage more students to pay attention to and learn BIM. In addition, BIM can be effectively linked with industry practice, such as carrying out practical cooperation projects with construction enterprises, so as to provide more practical opportunities and career orientation for students.
3. When formulating training programs and teaching BIM courses, universities should encourage students to participate in BIM competitions to improve their knowledge and application of BIM. Meanwhile, the training courses should emphasize the ease of use and usefulness of BIM to improve student experience. Educators are also suggested to provide more practical opportunities in the training process to further improve students' acceptance of BIM-related positions.

In terms of encouraging students to participate, due to the small proportion of students who can participate in high-level competitions, universities can actively organize school-level competitions, expand the coverage of BIM competitions by carrying out school-level college students' modeling, drawing and other competitions. At the same time, it can also carry out on-campus selection competitions for high-level BIM competitions, so that more students can participate in various innovative practice competitions related to BIM (S Wang, 2018). Van Nuland et al. (2015) finally verified that add competitions in teaching activities would significantly improve students' grades. Therefore, teachers who teach BIM technology can also release the assignments of actual engineering projects in the course, simulate the competition mechanism for students to complete in groups and stages, and the competition results are used as a part of the final grade of the course, so that to improve students' enthusiasm for learning BIM technology, and let more students join in BIM related competitions.

8.6.3 Future Work

In the next stage of the study, the dynamic relationship between BIM education and BIM-oriented career development in AEC disciplines will be explored in depth. A longitudinal dynamic tracking study of BIM competitions will be conducted to understand the changes in career competence and career planning before and after the competitions. In addition, the synergy between BIM education and industry practice, and the demand of the BIM industry for BIM technical talents can be explored. Further work can be performed on how to combine BIM education and industry practice to promote students' career development and industry development, and provide students with a more variety of options for career development.

References

Abernathy, T. V., & Vineyard, R. N. (2001). Academic competitions in science: What are the rewards for students? *The Clearing House: A Journal of Educational Strategies, Issues and Ideas, 74*(5), 269–276. https://doi.org/10.1080/00098650109599206

Agenda, I. (2016). *Shaping the future of construction a breakthrough in mindset and technology*. World Economic Forum.

Ahn, E., & Kim, M. (2016). BIM awareness and acceptance by architecture students in Asia. *Journal of Asian Architecture and Building Engineering, 15*(3), 419–424. https://doi.org/10.3130/jaabe.15.419

Ao, Y., Liu, Y., Tan, L., Tan, L., Zhang, M., Feng, Q., … Martek, I. (2021). Factors driving BIM learning performance: Research on China's sixth national BIM graduation design innovation competition of colleges and universities. *Buildings, 11*(12). https://doi.org/10.3390/buildings11120616

Ao, Y., Peng, P., Li, J., Li, M., Bahmani, H., & Wang, T. (2022). What determines BIM competition results of undergraduate students in the architecture, engineering and construction industry? *Behavioral Sciences, 12*(10). https://doi.org/10.3390/bs12100360

Bosnjak, M., Ajzen, I., & Schmidt, P. (2020). The theory of planned behavior: Selected recent advances and applications. *Europe's Journal of Psychology, 16*(3), 352–356. https://doi.org/10.5964/ejop.v16i3.3107

Burke, M. G., Carter, J. D., & Hughey, A. W. J. I. (2013). The use of case study competitions to prepare students for the world of work. *Industry and Higher Education, 27*, 157–162.

Casasayas, O., Hosseini, M. R., Edwards, D. J., Shuchi, S., & Chowdhury, M. (2021). Integrating BIM in higher education programs: Barriers and remedial solutions in Australia. *Journal of Architectural Engineering, 27*(1), 05020010. https://doi.org/10.1061/(ASCE)AE.1943-5568.0000444

Chen, S., & Chen, M. (2020). Research on the influencing factors of college students' participation in subject competition. *Economic Outlook the Bohai Sea, 07*, 141–143.

Damnjanovic, V., & Mijatovic, I. (2017). Student perception of benefits from being engaged in international case study competitions. *Management: Journal of Sustainable Business and Management Solutions in Emerging Economies, 22*(2), 61–72. https://doi.org/10.7595/management.fon.2017.0017

Davis, F. D., Bagozzi, R. P., & Warshaw, P. R. (1989). User acceptance of computer technology: A comparison of two theoretical models. *Management Science, 35*(8), 982–1003.

Doe, R., Kaur, K., Selway, M., & Stumptner, M. (2022). Interoperability in Aeco and the Oil & Gas Sectors: Object-Based Standards and Systems. *Journal of Information Technology in Construction, 27*, 312–334.

Donald, W. E., Ashleigh, M. J., & Baruch, Y. (2018). Students' perceptions of education and employability. *Career Development International, 23*(5), 513–540. https://doi.org/10.1108/cdi-09-2017-0171

Elgewely, M. H., Nadim, W., ElKassed, A., Yehiah, M., Talaat, M. A., & Abdennadher, S. (2021). Immersive construction detailing education: building information modeling (BIM)–based virtual reality (VR). *Open House International, 46*(3), 359–375. https://doi.org/10.1108/OHI-02-2021-0032

Elmualim, A., & Gilder, J. (2014). BIM: Innovation in design management, influence and challenges of implementation. *Architectural Engineering and Design Management, 10*(3–4), 183–199.

Fornell, C., & Larcker, D. F. (1981). Evaluating structural equation models with unobservable variables and measurement error. *Journal of Marketing Research, 18*, 39–50.

Fu, H., Xia, Z., Tan, Y., & Guo, X. (2024). Influence of cues on the safety Hazard recognition of construction workers during safety training: Evidence from an eye-tracking experiment. *Journal of Civil Engineering Education, 150*(1), 04023009. https://doi.org/10.1061/JCEECD.EIENG-1882

Grove, J. M. (2009). Cities: Managing densely settled social-ecological systems. In C. Folke, G. P. Kofinas, & F. S. Chapin (Eds.), *Principles of ecosystem stewardship: Resilience-based natural resource management in a changing world* (pp. 281–294). Springer. https://doi.org/10.1007/978-0-387-73033-2_13

Gu, T., Chen, Y., & Xiong, Y. (2021). Research on the construction of teaching satisfaction model of school-enterprise cooperative education-based on analysis of PLS-SEM. *Vocational Technology, 20*(06), 7–14.

Hair, J. F., Sarstedt, M., Ringle, C. M., & Mena, J. A. (2012). An assessment of the use of partial least squares structural equation modeling in marketing research. *Journal of the Academy of Marketing Science, 3*(40), 414–433.

Herrera-Limones, R., Rey-Pérez, J., Hernández-Valencia, M., & Roa-Fernández, J. J. S. (2020). Student competitions as a learning method with a sustainable focus in higher education: The University of Seville “Aura projects” in the “solar decathlon 2019”. *Sustainability, 12*, 1634.

Hong, S.-H., Lee, S.-K., Kim, I.-H., & Yu, J.-H. (2019). Acceptance model for mobile building information modeling (BIM). *Applied Sciences, 9*(18). https://doi.org/10.3390/app9183668

Jin, R., Yang, T., Piroozfar, P., Kang, B.-G., Wanatowski, D., Hancock, C. M., & Tang, L. (2018). Project-based pedagogy in interdisciplinary building design adopting BIM. *Engineering, Construction and Architectural Management, 25*(10), 1376–1397. https://doi.org/10.1108/ecam-07-2017-0119

Jin, R., Zou, P. X., Li, B., Piroozfar, P., & Painting, N. (2019). Comparisons of students' perceptions on BIM practice among Australia, China and UK. *Engineering, Construction and Architectural Management, 26*(9), 1899–1923. https://doi.org/10.1108/ecam-07-2018-0275

Li, J. (2021). Discussion on graduation design reform of engineering management specialty based on BIM. *Science & Technology Information, 19*(32), 112–114. https://doi.org/10.16661/j.cnki.1672-3791.2111-5042-6064

Li, J., Afsari, K., Li, N., Peng, J., Wu, Z., & Cui, H. (2020). A review for presenting building information modeling education and research in China. *Journal of Cleaner Production, 259*, 120885.

Miettinen, R., & Paavola, S. (2014). Beyond the BIM utopia: Approaches to the development and implementation of building information modeling. *Automation in Construction, 43*, 84–91.

Nießen, D., Wicht, A., Schoon, I., & Lechner, C. M. (2022). “You can't always get what you want”: Prevalence, magnitude, and predictors of the aspiration–attainment gap after the school-to-work transition. *Contemporary Educational Psychology, 71*. https://doi.org/10.1016/j.cedpsych.2022.102091

Okay-Somerville, B., & Scholarios, D. (2014). Coping with career boundaries and boundary-crossing in the graduate labour market. *Career Development International, 19*(6), 668–682. https://doi.org/10.1108/cdi-12-2013-0144

Ou, Y., Li, K., & Zhao, Y. (2014). BIM, second architectural design revolution. *Architecture Technique*, (02), 26–29.

Peng, P., Ao, Y., Li, M., Wang, Y., Wang, T., & Bahmani, H. (2022). Building information modeling learning behavior of AEC undergraduate students in China. *Behavioral Sciences, 12*(8). https://doi.org/10.3390/bs12080269

Plakhotnik, M. S., Krylova, A. V., & Maslikova, A. D. (2020). Does participation in case competitions improve career decision-making self-efficacy of university students? *Education + Training, 62*(6), 659–675. https://doi.org/10.1108/et-01-2020-0021

Pynoo, B., Tondeur, J., van Braak, J., Duyck, W., Sijnave, B., & Duyck, P. (2012). Teachers' acceptance and use of an educational portal. *Computers & Education, 58*(4), 1308–1317. https://doi.org/10.1016/j.compedu.2011.12.026

Qin, X., Shi, Y., Lyu, K., & Mo, Y. (2020). Using a tam-toe model to explore factors of building information modelling (Bim) adoption in the construction industry. *Journal of Civil Engineering and Management, 26*(3), 259–277. https://doi.org/10.3846/jcem.2020.12176

Sackey, E., Tuuli, M., & Dainty, A. (2015). Sociotechnical systems approach to BIM implementation in a multidisciplinary construction context. *Journal of Management in Engineering, 31*(1), A4014005.

Sacks, R., & Pikas, E. (2013). Building information modeling education for construction engineering and management. I: Industry requirements, state of the art, and gap analysis. *Journal of Construction Engineering and Management, 139*(11), 04013016.

Sadan, K. -K. (2021, November 12). *Overview of student innovation competitions and their roles in STEM education.* Virtually hosted by the section https://peer.asee.org/38445

Sayani, H., Shoaib, M., Mashood, N., & Kumar, V. (2019). Case study competitions and perceptions of learning among undergraduate students in The United Arab Emirates. *Journal of Education for Business, 94*(2), 101–112. https://doi.org/10.1080/08832323.2018.1503581

Schmitt, N. (1996). Uses and abuses of coefficient alpha. *Psychological Assessment, 4*(8), 350–353.

Sev, A. (2009). How can the construction industry contribute to sustainable development? A conceptual framework. *Sustainable Development, 17*(3), 161–173.

Sotelino, E. D., Natividade, V., & Travassos do Carmo, C. S. (2020). Teaching BIM and its impact on young professionals. *Journal of Civil Engineering Education, 146*(4), 05020005. https://doi.org/10.1061/(asce)ei.2643-9115.0000019

Sun, Y., & Cao, S. (2021). Factors of willingness to learn BIM technology in college students based on technology acceptance model. *Project Management Technology, 19*(3), 70–76.

Tang, C., Han, Y., & Ng, P. (2022). Green consumption intention and behavior of tourists in urban and rural destinations. *Journal of Environmental Planning and Management*, 1–25. https://doi.org/10.1080/09640568.2022.2061927

Tsai, M.-H., Chen, K.-L., & Chang, Y.-L. (2019). Development of a project-based online course for BIM learning. *Sustainability, 11*(20). https://doi.org/10.3390/su11205772

Van Nuland, S. E., Roach, V. A., Wilson, T. D., & Belliveau, D. J. (2015). Head to head: The role of academic competition in undergraduate anatomical education. *Anatomical Sciences Education, 8*(5), 404–412. https://doi.org/10.1002/ase.1498

Vigneshwar, R. V. K., Shanmugapriya, S., & Sindhu Vaardini, U. (2022). Analyzing the driving factors of BIM adoption based on the perception of the practitioners in Indian construction projects. *Iranian Journal of Science and Technology, Transactions of Civil Engineering, 46*(3), 2637–2648. https://doi.org/10.1007/s40996-022-00834-9

Wang, S. (2018). The applied undergraduate civil engineering major is based on BIM practical teaching research of design competition mechanism. *Information Records and Materials, 19*, 140–141.

Wang, H., & Meng, X. (2021). BIM-supported knowledge management: Potentials and expectations. *Journal of Management in Engineering, 37*(4), 04021032.

Wang, L., Huang, M., Zhang, X., Jin, R., & Yang, T. (2020). Review of BIM adoption in the higher education of AEC disciplines. *Journal of Civil Engineering Education, 146*(3). https://doi.org/10.1061/(asce)ei.2643-9115.0000018

Wankat, P. C. (2005). Undergraduate student competitions. *Journal of Engineering Education, 94*(3), 343–347. https://doi.org/10.1002/j.2168-9830.2005.tb00860.x

Wu, Y. (2018). Analysis of the influence of skill competition on the career development of higher vocational college graduates. Journal of Jiyuan Vocational and Technical College, 17(03), 45–49.

Wu, W., & Issa, R. R. A. (2014). BIM education and recruiting: Survey-based comparative analysis of issues, perceptions, and collaboration opportunities. *Journal of Professional Issues in Engineering Education and Practice, 140*(2), 04013014. https://doi.org/10.1061/(asce)ei.1943-5541.0000186

Wu, W., Mayo, G., McCuen, T. L., Issa, R. R., & Smith, D. K. (2018). Building information modeling body of knowledge. II: Consensus building and use cases. *Journal of Construction Engineering and Management, 144*(8), 04018066.

Wu, Z., Jiang, M., Li, H., Luo, X., & Li, X. (2021). Investigating the critical factors of professionals' BIM adoption behavior based on the theory of planned behavior. *International Journal of Environmental Research and Public Health, 18*(6), 3022. https://doi.org/10.3390/ijerph18063022

Yang, Y., & Ha, M. (2018). Study on the influence factors of teaching quality of BIM in the mode of promoting teaching and learning by competition. *Journal of Hebei University of Engineering (Natural Science Edition), 35*(3), 106–109. https://doi.org/10.3969/j.issn.1673-9469.2018.03.021

Zhang, L., Chu, Z., & Song, H. (2019). Understanding the relation between BIM application behavior and sustainable construction: A case study in China. *Sustainability, 12*(1). https://doi.org/10.3390/su12010306

Zhao, Y., Sun, Y., Zhou, Q., Cui, C., & Liu, Y. (2022). How AEC professionals accept BIM technologies in China: A technology acceptance model perspective. *Engineering, Construction and Architectural Management*, (ahead-of-print). https://doi.org/10.1108/ecam-04-2022-0308

Chapter 9
Conclusion and Future Research: Bridging BIM Education, Engagement, and Industry Readiness

Abstract This chapter summarizes the main findings of the book and outlines directions for future research. The study constructs and validates an integrated framework linking "learning motivation–behavioral intention–capacity building–career orientation," revealing how classroom learning, competition participation, and psychological mechanisms jointly shape educational outcomes and career orientation. It addresses the theoretical gap in the "classroom–competition–employment" chain. The research also compares students and industry professionals in terms of educational needs and behavioral mechanisms, and develops a multidimensional evaluation system for BIM education outcomes, providing references for curriculum optimization and talent cultivation. Limitations mainly lie in the reliance on cross-sectional data, lack of objective behavioral evidence, and limited sample representativeness. Future research should emphasize longitudinal tracking, multi-context experiments, and cross-cultural comparisons, while exploring the integration of BIM education with emerging technologies such as AI and VR/AR, in order to strengthen education–industry alignment and advance international development.

Keywords BIM education · Learning motivation · Competition participation · Career orientation · Educational outcomes

9.1 Research Contributions

This book focuses on exploring the learning behaviors of Chinese university students in architecture, engineering, and construction (AEC) disciplines during the BIM education process, the mechanisms of their participation in competitions, and the pathways through which these factors influence career development. The study offers four distinctive contributions and innovations:

1. Filling the gap in BIM education behavior research and promoting alignment between education and industry competency requirements. This research centers

Y. Ao et al., *Building Information Modeling (BIM) Adoption*, Resilient and Friendly Living Environments, https://doi.org/10.1007/978-981-95-3993-2_9

on the BIM learning behaviors and competition participation mechanisms of AEC students in Chinese universities, constructing an integrated empirical framework that spans from classroom learning to competition practice and ultimately to career intentions. It addresses the current theoretical gap in the "classroom–competition–employment" continuum within BIM education research. Compared to prior studies that have focused mainly on developing teaching resources or building course systems (often concentrating on curriculum structure), this study adopts the student perspective, conducting an in-depth examination of their learning motivations, behavioral patterns, and career focus pathways. The findings offer a new research paradigm for BIM talent cultivation. Moreover, the results help educational authorities and industry employers better understand students' actual learning behaviors and needs, thereby fostering more precise alignment between BIM teaching content in universities and the skill requirements of the industry, ultimately enhancing graduates' industry readiness.

2. Revealing the intrinsic mechanism linking learning motivation, behavioral intention, and career orientation. Drawing on the Unified Theory of Acceptance and Use of Technology (UTAUT), the Theory of Planned Behavior (TPB), and Social Cognitive Theory (SCT), this study innovatively constructs a behavioral model of students' BIM learning and competition participation. Using Structural Equation Modeling (SEM), the research clarifies the influence pathways of psychological factors—such as attitude, self-efficacy, and social influence—on students' behavioral intentions, and reveals how these factors subsequently affect their actual participation in BIM competitions. The findings indicate that competition experience has a significant positive effect on students' subsequent interest in BIM-related careers. This demonstrates the inherent linkage between informal learning experiences (such as academic competitions) and career choice—competition participation can strengthen students' confidence in and interest toward pursuing a career in BIM. This finding enriches theoretical understanding in the field of educational behavior regarding the relationship between learning motivation and career orientation, and provides empirical support for using competitions and other practice-based teaching methods to stimulate students' professional interest.
3. Developing an evaluation system for BIM education outcomes in competition-based learning contexts. Addressing the limitations of traditional BIM teaching assessments—which rely heavily on student self-reports and theoretical exam scores—this study, based on large-scale surveys, proposes a multidimensional BIM education effectiveness index system. The framework covers key dimensions such as teacher support, team collaboration, individual skill enhancement, and industry awareness, enabling a comprehensive reflection of the educational outcomes produced by competition-based learning. Through factor analysis and regression modeling, the validity and predictive power of the framework have been verified, demonstrating its applicability to classroom teaching evaluation, competition result assessment, and student growth path analysis. This evaluation framework enriches the theoretical and methodological toolkit for BIM teaching assessment, allowing educators to more scientifically measure the value of teach-

ing practices and competition activities, and providing data-driven support for the optimization of instructional strategies.

4. Incorporating student psychological factors to reinforce a human-centered perspective in BIM education. This research incorporates psychological variables such as behavioral intention, self-efficacy, and social identity into the analysis, examining students' BIM learning processes through a full "cognition–affection–behavior" chain perspective. The study explores students' subjective perceptions of BIM learning difficulty, their attitudes toward competition-related stress, and the sources of their intrinsic motivation, finding that these psychological factors play a pivotal role in learning engagement and competition performance. Compared to traditional teaching research focused mainly on "tools and skill mastery," this study offers a student-centered perspective that integrates both psychological experience and developmental needs. The introduction of this human-centered lens not only deepens the theoretical contribution of the research but also provides important insights for constructing BIM teaching models with greater humanistic care—emphasizing that alongside technical competence development, greater attention should be paid to students' psychological well-being, career confidence, and long-term growth.

9.2 Research Limitations

Constrained by data acquisition conditions, questionnaire design restrictions, and the coverage scope of competition platforms, this study inevitably has several limitations:

1. Data collection is primarily cross-sectional, making it difficult to fully capture the dynamic evolution of student behaviors. The data for this research is mainly drawn from questionnaire surveys conducted at specific time points. Although the sample covers students from different regions, grades, and majors across China, it lacks longitudinal tracking of the same cohort over time. This cross-sectional design limits our ability to observe changes in students' BIM learning motivations and strategies at different stages, as well as the long-term effects of educational interventions on capability development. Consequently, the findings reflect characteristics of the survey period rather than long-term dynamic patterns.
2. Evaluation indicators for educational outcomes are largely based on subjective perceptions, lacking support from objective behavioral data. In assessing teaching effectiveness, we employed structured questionnaires and Likert scales; however, all data was self-reported by students, inevitably subject to cognitive bias or emotional influence. For instance, some students may overestimate or underestimate their skill improvement, and perceptions of instructional support may vary depending on personal expectations. This means our evaluation results lack corroboration from objective behavioral data—such as the quality of com-

pleted BIM modeling tasks, records of team collaboration, or competition platform activity logs—thereby leaving room for improvement in the external validity of the conclusions.

3. Competition samples are concentrated on a single event platform, limiting the generalizability of conclusions. The primary sample for this study comes from participants in the National BIM Graduation Design Innovation Competition. Since its inception in 2013, this event has grown in scale and influence, and was listed as a key university competition by the Ministry of Education in 2020. While it enjoys high authority and broad participation, its organization model, scoring criteria, and task design have certain unique features that may cause some findings to be event-specific. Variations exist among Chinese regions and universities in the extent and form of BIM competition participation, meaning the applicability of our conclusions to broader competition contexts requires further testing. For example, applying our analytical framework to other BIM-related competition platforms or to practice-based teaching activities not tied to competitions could help determine whether our conclusions hold more generally.
4. The research sample is limited to mainland Chinese universities, lacking cross-regional and international comparative perspectives. Although our survey covers universities in all seven major regions of mainland China (East, Central, and West), it does not include data from Hong Kong, Macao, Taiwan, or overseas institutions. Differences in educational systems, cultural contexts, and industry maturity may significantly shape BIM education models and student behaviors. Without cross-regional or international comparative analysis, this study cannot fully assess the local characteristics and general patterns of BIM education in China, nor can it verify the applicability of our theoretical model in other cultural contexts. This limitation suggests caution when attempting to generalize the findings internationally and highlights the need for more diverse samples in future research to strengthen the generalizability of results.
5. Some variables influencing competition outcomes have not been explored in depth, leaving mechanisms insufficiently analyzed. Although we constructed a path model from students' behavioral motivations to competition participation outcomes, certain potentially important variables were not quantified in detail. For example, in team competition settings, the impact of different role allocations on learning outcomes, the challenges posed by varying task complexity to skill application, and the guiding role of judges' criteria awareness on participation strategies were not sufficiently incorporated into the model. These nuanced factors in competition contexts may significantly affect the benefits of competition-based learning, yet the study lacked specifically designed indicators and structures to analyze them in depth. Consequently, our understanding of the internal operational mechanisms of BIM competitions remains incomplete and would benefit from further elaboration.

Despite these limitations, this book represents the first empirical study in the Chinese context to systematically focus on the "learning–competition–employment" chain in BIM education behavior, providing a solid foundation for more in-depth and broader research in the future. These limitations do not diminish the theoretical value or methodological significance of the study; rather, they point toward concrete directions for continuous improvement.

9.3 Future Prospects

In response to the research limitations outlined above, the following recommendations are proposed for future studies, providing reference and direction for subsequent scholars:

1. Adopt multiple data sources to capture the full process of student learning behaviors. To more precisely depict BIM learning behaviors, future research should incorporate diverse data collection methods, integrating traditional surveys with learning analytics techniques. On one hand, online BIM platforms or software can be used to automatically log students' learning process data—such as modeling operation records, team collaboration communications, and version histories of submitted work—to capture skill acquisition and collaboration dynamics at a micro level. On the other hand, mobile applications and real-time feedback tools can collect students' immediate experiences and reactions across various learning scenarios. The integration of multiple data sources will help overcome the limitations of questionnaire-only approaches, comprehensively mapping students' learning pathways and behavioral evolution, and providing evidence for optimizing instructional interventions.
2. Conduct longitudinal tracking to reveal the long-term impact of educational interventions. It is recommended to track specific student cohorts over an extended period (e.g., 2–4 years), observing their progression from introductory BIM courses through competition practice to post-graduation employment. By regularly collecting behavioral and psychological data from the same group at key stages—course learning, competition participation, internships, and job placement—a time-series dataset can be constructed to analyze the long-term effects of educational interventions and competition experience on skill development and career choices. For instance, comparing BIM skill levels and career outcomes of competition participants versus non-participants at graduation can verify the delayed effects of competition-based learning. Longitudinal studies can reveal dynamic change patterns, address the shortcomings of cross-sectional designs, and provide educators with more forward-looking decision-making support.
3. Design multi-context experiments and task simulations to test teaching strategy effectiveness. Given that BIM teaching and competitions are inherently task-driven learning modes, future research can introduce controlled variables into experimental teaching environments to compare instructional designs. For example, different classes or teams can be assigned varying task complexities, team configurations, or feedback frequencies, simulating multiple teaching con-

texts to examine the effects on student engagement, team performance, and knowledge transfer. Through experimental and control group comparisons, researchers can rigorously verify which instructional strategies most effectively enhance BIM skills and innovation capacity. Such context-based simulation studies can provide evidence-based practical guidance for BIM education reform, helping universities design more effective courses and competition training programs.
4. Expand the regional and cultural scope of research to conduct cross-cultural comparisons. To explore both the general patterns and unique characteristics of BIM education, future studies should incorporate more diverse sample backgrounds. On one hand, research could be extended to universities in Hong Kong, Macao, and Taiwan to examine similarities and differences in BIM teaching under different education systems. On the other hand, comparative studies could be conducted with overseas universities in "Belt and Road" countries or other regions with active BIM education practices, analyzing how cultural differences and varying levels of industrial and technological maturity shape students' BIM learning behaviors and influencing factors. Cross-regional and cross-cultural analyses could distill universally applicable BIM education mechanisms and competency assessment standards, promote the international applicability of China's experiences, and enrich the global research landscape on BIM education.
5. Promote the integration of BIM education with emerging digital technologies to explore innovative teaching models. In the face of the digital transformation wave, BIM education should actively embrace the integration and empowerment of next-generation information technologies. Future studies could focus on teaching models that deeply integrate BIM with artificial intelligence (AI), virtual/augmented reality (VR/AR), and digital twin technologies. For example, VR/AR could be used to create immersive virtual construction environments, enabling students to engage in BIM practice in simulated engineering contexts; AI-assisted teaching systems could provide real-time personalized guidance and feedback based on students' learning behavior data; and digital twin technologies could link real-world projects to classroom teaching, enhancing students' awareness of and ability to address practical engineering challenges. These innovations have the potential to improve teaching experiences and learning outcomes, cultivating BIM talent better equipped for the era of smart construction.

In summary, as the first empirical study in the Chinese context to systematically focus on the complete chain of "classroom learning–competition practice–career orientation" in BIM education behavior, this research has provided valuable theoretical support and data-driven insights for understanding behavioral mechanisms, optimizing instructional strategies, and strengthening education–industry alignment. Although still at an exploratory stage, its findings and contributions lay a solid foundation for more extensive and in-depth studies in the future. By continually refining research methods and broadening perspectives, we will be able to further capture students' genuine needs and capacity growth patterns, advancing BIM education in universities toward a more systematic, evidence-based, and industry-synchronized new stage.

Appendix 1: Questionnaire on BIM Learning Behavior of Chinese AEC Undergraduates

Performance expectancy:

PE1—Learning BIM technology is helpful for my studies.
PE2—Learning BIM technology is helpful for my career development.
PE3—Learning BIM technology can make me more competitive when I enter the workplace in the future.

Effort expectancy:

EE1—I think BIM software interface design is reasonable and easy to operate.
EE2—I think it's easy to learn to use BIM software.
EE3—It's easy for me to be proficient in using BIM software.

Social influence:

SI1—People who are important to me think I should learn BIM technology.
SI2—People who influence my behavior think I should learn BIM technology.
SI3—The promotion and popularization of BIM technology in the construction industry prompted me to learn BIM.

Facilitating conditions:

FC1—The school supports us to learn BIM technology.
FC2—The BIM instructor can give me useful suggestions.
FC3—I have the resources needed to learn BIM technology.
FC4—When I encounter difficulties in learning BIM technology, teachers/classmates can help me solve the difficulties.

Learning attitude:

LA1—It is worthwhile to learn BIM technology.
LA2—I am interested in BIM technology.
LA3—It's fun to learn BIM technology.

Y. Ao et al., *Building Information Modeling (BIM) Adoption*, Resilient and Friendly Living Environments, https://doi.org/10.1007/978-981-95-3993-2

Learning intention:

LI1—I want to continue learning BIM technology in the future.
LI2—I want to work related to BIM in the future.
LI3—I'd like to take part in the BIM competition.

Learning behavior:

LB1—I can operate related BIM software.
LB2—I am learning or have already learned BIM technology.

Appendix 2: BIM Education Effectiveness Questionnaire

Factor category	Latent variable	Code	Items
Environment factors	BIM theory education environment (TE)	TE1	Through the theoretical courses at the university, I have acquired a good grasp of the professional skills required to work in BIM technology-related jobs (software modeling, 5D construction simulation, etc.)
		TE2	The school I attended had a good education system for BIM technology
		TE3	The instructors of courses related to BIM technology are highly specialized and attach great importance to the lectures.
		TE4	Through the theoretical courses at the university, I have a clear perception of the job positions related to BIM technology
		TE5	I think the incentives for BIM technology education in schools are comprehensive
		TE6	I think the school's BIM vocational education incentives work well
	BIM practice education quality (PQ)	PQ1	High-quality completion of BIM technology-related course design and participation in BIM technology-based skill competitions have a positive impact on professional perception and confidence in BIM practice
		PQ2	High-quality completion of BIM technology-related course design and participation in BIM technology skills competitions have a positive impact on employment and career development
		PQ3	Construction companies fully recognize the role of BIM skills competitions and BIM technology-related course designs
		PQ4	The school's BIM technology program design and skills competitions are highly relevant to my future job

Y. Ao et al., *Building Information Modeling (BIM) Adoption*, Resilient and Friendly Living Environments, https://doi.org/10.1007/978-981-95-3993-2

Factor category	Latent variable	Code	Items
Cognitive factors	BIM career decision self-efficacy	SE1	I am aware that my personality is suitable for a BIM-related job position
		SE2	I am certain that a BIM-related job position would best utilize my skills
		SE3	I think a BIM-related job is the perfect career for me
		SE4	I would like to learn more about BIM technology.
		SE5	I found that my skills, values, and interests were a good match for BIM-related work
		SE6	For BIM-related career choices, I can make an informed choice
		SE7	I am willing to learn more about BIM technology-related knowledge
		SE8	I'm sure I'm interested in BIM-related work.
	BIM career decision coping efficacy	CE1	If my first choice for a BIM-related job position doesn't work out, I can be honest about it
		CE2	If my first choice for a BIM-related job position doesn't work out, I'll find another career choice and manage the negativity
		CE3	If I find that I am unable to perform BIM-related work due to subjective reasons (lack of my skills, etc.), I will develop an alternative plan
		CE4	If I find that I can't do a BIM-related job for objective reasons (downturn in the construction industry, etc.), I'll develop an alternative plan
Behavioral factors	BIM education effectiveness	EE1	I've been keeping up to date with the latest news and developments in BIM-related work.
		EE2	I took the initiative to participate in BIM technology skill competitions and learn related skills
		EE3	I focused on BIM-related job career prospects and trends
		EE4	I plan to participate in BIM-related job openings and deepen my skills in BIM-related technologies

Structural equation modeling total effects				
Path	Original sample(0)	Standard deviation(STDEV)	T statistics(|0/STDEV|)	P values
CDCE - > EE	0.182	0.053	3.404	0.001(***)
CDSE - > EE	0.707	0.053	13.331	0.000(***)
TE - > CDCE	0.124	0.061	2.024	0.043(*)
TE - > CDSE	0.403	0.053	7.544	0.000(***)
TE - > EE	0.338	0.070	4.802	0.000(***)
PQ - > CDCE	0.452	0.061	7.465	0.000(***)
PQ - > CDSE	0.324	0.057	5.727	0.000(***)
PQ - > EE	0.218	0.069	3.14	0.002(**)

Appendix 3: BIM Competition Intention Questionnaire

Performance Expectancy (PE):

PE1—Participating in a BIM competition can improve your basic professional skills.
PE2—Participating in a BIM competition can improve teamwork and communication skills.
PE3—Participation in BIM competition can enrich the experience of BIM software operation, which is good for finding a job after graduation.

Effort Expectancy (EE):

EE1—I think BIM software is easy to learn.
EE2—I think it is easy to participate in the competition using BIM software.
EE3—I think it is easy to get a good result in the BIM competition.

Social Influence (SI):

SI1—People who influence my behavior think I should learn BIM technology.
SI2—The people I care about think I should enter the BIM competition.
SI3— People who influence my behavior think I should enter BIM competitions.

Attitude (AT):

AT1—I think it is worthwhile to participate in the BIM competition.
AT2—It was fun to participate in the BIM competition.
AT3—I would like to participate in the BIM competition

Facilitating Conditions (FC):

FC1—The school supports us in learning BIM technology.
FC2—The school/institute supported us to participate in the BIM competition

Y. Ao et al., *Building Information Modeling (BIM) Adoption*, Resilient and Friendly Living Environments, https://doi.org/10.1007/978-981-95-3993-2

Self-efficacy (SE):

SE1—As long as I do my best, I can always solve the problems I encounter in the BIM competition.
SE2—It was easy for me to stay in the game and achieve my goal.

Behavioral Intention (BI):

BI1—I want to learn more about BIM technology in the future.
BI2—I would like to participate in the BIM competition.

Student Behavior (SB):

SB1—Among the various competitions, I pay more attention to the BIM category.

Appendix 4: BIM Competition Experience Questionnaire

Perceived Ease of Use (PEU):

PEU1—I think the BIM software interface is well designed and easy to use.
PEU2—I think the concept of BIM is clear and understandable.
PEU3—I think the BIM competition rules and procedures are clear and understandable.

Perceived Usefulness (PU):

PU1—Learning BIM will help me in my career development.
PU2—Participating in a BIM competition can improve professional skills.
PU3—Participating in BIM competitions can improve teamwork and communication skills.
PU4—Participating in BIM competition can enrich the experience of BIM software operation and help you find a job after graduation.

Subjective Norm (SN):

SN1—People who influence my behavior think I should learn BIM.
SN2—People I care about think I should participate in BIM competitions.
SN3—People who influence my behavior think I should participate in BIM competitions.

Perceived Behavioral Control (PBC):

PBC1—I have the resources to learn BIM.
PBC2—I have the basic professional knowledge needed to learn BIM.
PBC3—As long as I try my best, I can always solve the problems I encounter in the BIM competition.

Y. Ao et al., *Building Information Modeling (BIM) Adoption*, Resilient and Friendly Living Environments, https://doi.org/10.1007/978-981-95-3993-2

Attitude (AT):

AT1—Learning BIM is fun.
AT2—Participating in BIM competitions is fun.
AT3—I enjoy participating in BIM competitions.

Behavioral Intention (BI):

BI1—I want to participate in BIM competitions.
BI2—I plan to continue learning or using BIM.
BI3—I want to work in BIM-related jobs in the future.

Student Behavior (SB):

SB1—I pay more attention to BIM competitions among all types of competitions.
SB2—I pay more attention to BIM jobs when looking for part-time internships and career choices.

MIX
Papier aus verantwortungsvollen Quellen
Paper from responsible sources
FSC® C105338

If you have any concerns about our products,
you can contact us on
ProductSafety@springernature.com

In case Publisher is established outside the EU,
the EU authorized representative is:
Springer Nature Customer Service Center GmbH
Europaplatz 3, 69115 Heidelberg, Germany

Printed by Libri Plureos GmbH
in Hamburg, Germany